Atlas of the Sea

Illustrated by
David Nockels
Consultant Sir George Deacon

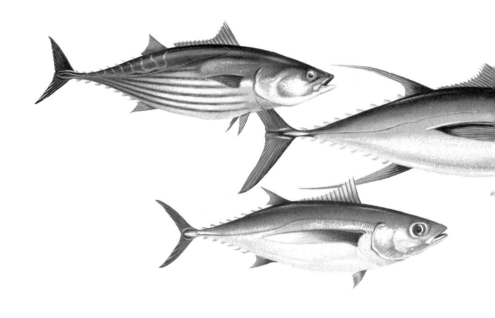

Maps by Geographical Projects London

Atlas of the Sea

Robert Barton

Heinemann London

Geographical Director **Shirley Carpenter**
Editor **Geoffrey Rogers**
Design **Roger Hyde**
Douglas Sneddon

William Heinemann Ltd., 15–16 Queen Street, London, W1X 8BE
London   Melbourne   Toronto
Johannesburg   Auckland

First Published 1974
© Aldus Books Limited, London, 1974
SBN 434 04800 3
Printed and bound in Spain by Novograph S.L. and Roner S.A.
Crta. de Irun Km. 12,450
Madrid 34
Dep. Legal: M. 23.683-1974.

Contents

		Introduction	7
Chapter	1	One Ocean	8
	2	The Atlantic Ocean	16
	3	The Pacific Ocean	32
	4	The Indian Ocean	48
	5	The Arctic Ocean	64
	6	The Southern Ocean	72
	7	Far Eastern Seas	84
	8	The Caribbean and Gulf of Mexico	92
	9	The Mediterranean Sea	102
	10	The North Sea	112
		Index	125

Introduction

In a century when six men have walked on the moon, only two have been to the deepest part of the ocean. The ocean is the last great area left on this earth to explore. It is vital to man's survival to discover the ocean's secrets, for the oceans play a huge part in controlling world climate; they have vast untapped resources of food, often close to areas where protein is desperately needed; there are chemicals and minerals—in solution, or lying on the seabed, or beneath it—to supply all man's future needs.

This book looks at the world ocean's mighty forces and resources. It discusses huge undersea mountain ranges and deep trenches, the full significance of which have only recently been appreciated and which have caused a fundamental re-thinking about the way the earth developed. It looks at the huge and silent "rivers" of the oceans—currents that sweep for thousands of miles round the earth, and which affect our lives every day from the food we eat to the clothes we wear and the houses we live in.

Chapters 2 to 6 look in turn at the "great" oceans, dwelling on the main features of each, including the undersea mountain ranges of the Atlantic; twelve-inch high waves that race across the Pacific at 500 miles an hour; the huge twice-yearly change of direction of Indian Ocean currents; man's quest across the Arctic either through the ice or below the surface; and his remorseless slaughter of whales in the Southern Ocean. The resources of the oceans are discussed, too, from the multinational North Atlantic fishing industry to ambitious plans to use huge "vacuum cleaners" to suck up mineral riches from the floor of the Pacific; from the world-ranging Japanese tuna fishery, to a plan to tow icebergs from the Southern Ocean to California to ease water shortages.

Chapters 7 to 10 turn their attention to the main seas of the world—the intricate pattern of Far Eastern Seas; the Caribbean and Gulf of Mexico; the Mediterranean; and the North Sea. Main features and resources are described—and a warning note is sounded about the limits to which man can go in his abuse of the ocean both by over-exploitation and pollution.

1 One Ocean

Man is a land-dwelling animal. He thinks of Planet Earth as land. Even the word "earth" has the twin meaning of "planet" and "soil" or "land." The sea is thought of as something that occurs between land masses. In fact land is something that occasionally obtrudes above the otherwise flat surface of the sea.

A creature from outer space coming to this planet purely by chance would have a totally different view of Earth to the one we hold. He would see Earth as a liquid planet. For a start, the chances of his landing in the sea are three times as great as they are of his arriving on land. And there is an almost even chance that he would sink down onto one of the great ocean's abyssal plains, since these take up nearly half the area of the earth, lying beneath 12,000 to 18,000 feet of water. Our explorer would find that these plains only loosely fit the name given to them. True, for great expanses they are flat and fairly featureless, covered with material that has rained down from above: sediment, the skeletons of fish and marine animals, the remains of ships, aircraft, and men. They are the graveyards of the oceans. But he would also find a "landscape" more dramatic than anything ashore: towering mountains rise to within 2,000 feet of the sea surface and sometimes pierce the surface to form mid-oceanic islands; and trenches plunge suddenly to depths of 20,000 feet, so that at their deepest parts they are 36,000 feet below the ocean surface.

As well as being graveyards, the abyssal plains support a wide variety of life. Worms make intricate patterns as they wind through the soft bottom sediments; snails and shrimps search blindly for food in the perpetual darkness; strangely shaped fish, some without eyes, some almost completely transparent,

Man has divided up the world ocean and imposed arbitrary boundaries. But it is one ocean, and the water that gently laps a coral atoll in the Pacific is the same as that which pounds the granite cliffs of Scotland. Life began in the oceans; the oceans hold the key to all future life on earth.

WORLD OCEANS
General Features

WORLD OCEANS
Surface Currents

some with weirdly flickering lights at the entrance to their large mouths to entice their prey, swim slowly through the cold water.

Crossing the abyssal plain our explorer would eventually find himself on a slope that rises fairly steeply—about one in ten—towards the surface. This is the continental slope, leading up to the continental shelf, and is the true boundary between the continental land masses and the ocean. If the sea was drained of its water the land areas would stand out on the continental shelves, high above the ocean basins. Drain off the water for about only 1,000 feet and it is easy to see how the British Isles, for example, are really part of the European continent, separated only by a shallow ditch of water.

On the continental shelf, which extends on average to a depth of about 600 feet, our visitor from outer space increasingly encounters signs of man, and even man himself. Clumsy-looking nets scrape along the seabed to scoop up fish; drill pipes reach down into the seabed; small submarines feel their way gingerly through the gloom; men in rubber suits struggle to perform simple tasks on the seabed. There are pipelines threading their way through the sand, structures rising out of the seabed towards the surface, and anchors and cables litter the underwater landscape.

If he stayed beneath the sea for any time our visitor would discover how little of the vast expanse of the oceans man has conquered. He would see that man cannot yet descend, at best, more than 1,000 feet below the surface of the sea without protective covering; that he has only a sketchy knowledge of its currents and tides; that he is only just beginning to sample its riches to any real extent, yet those riches that he has learnt to harvest he seems intent on ruthlessly destroying; that much of marine life is as yet unknown to him. Our visitor would discover, worst of all, that despite this lack of knowledge, man treats the oceans with contempt, using them as a dumping ground for anything he no longer has use for.

Emerging from the sea, our visitor would be exposed to an entirely different environment: one that baked or froze him with alarming speed over relatively short distances; the winds of which buffeted him; an environment that rained on him, hailed on him, and plunged him alternatively from bright light into inky darkness. He would probably quickly conclude that the sea is the logical home for an intelligent, reasoning creature. After all, it covers the greatest area of the planet (71 per cent). And it is far more equable than the land: it heats up and cools down slowly; its total temperature range is small compared to that of the land; its chemical composition varies hardly at all over its entire area; although its surface is often disturbed by winds, just below this surface all is calm, all is perpetually dark, and the effects of gravity (which man struggles at considerable expense to overcome on land) are greatly diminished.

On land, our visitor would see that man is rapidly outgrowing the 29 per cent of the earth's surface he has claimed as his own. Millions of people do not have enough to eat. Natural resources are being used up fast, yet demand is increasing at a frightening

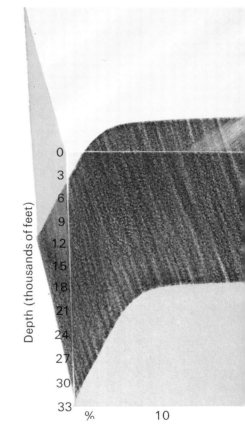

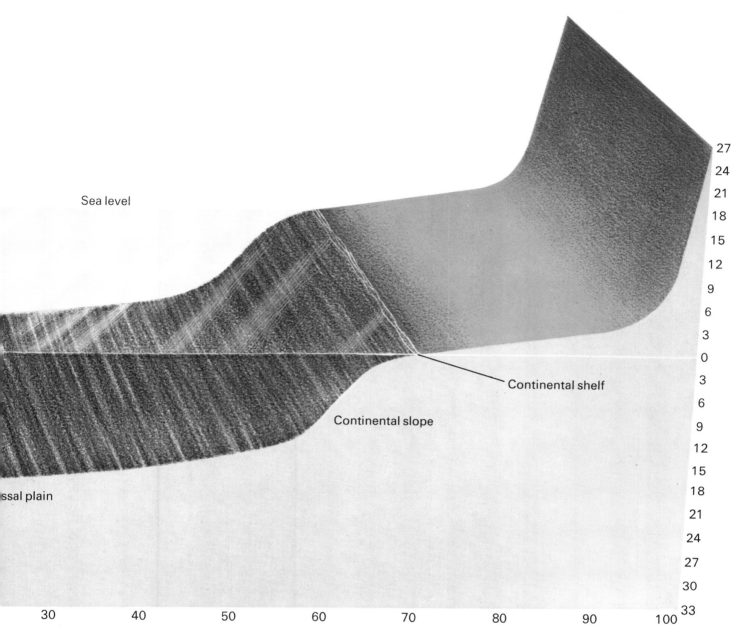

Just over 70 per cent of the earth's surface is covered by water, the larger part of which is 12,000 feet or more deep, as the diagram above shows. Between depths of 12,000 feet and 18,000 feet the seabed is known as the abyssal plain—rather a misleading name because it supports towering seabed mountains and is split by ocean trenches as deep as 36,000 feet. Approximately 20 per cent of the earth's surface lies under water between 600 and 12,000 feet deep. This is the continental slope, the true boundary between the continental land masses and the ocean. The continental shelf, which extends to an average depth of 600 feet and where fishing and other marine activities are most intense, forms only 5 per cent of the earth's surface.

rate. This means that man will more and more turn to the sea for food, for essential minerals, for water, even for living space. This book looks at the world oceans and at those seas where man has made the greatest strides in tapping oceanic resources. What it is hoped will emerge is some idea of the sheer size of the oceans and the immense difficulties inherent in man's attempts to learn more about them and to harness their wealth. Each chapter deals with one or more major oceanic subject, ranging from continental drift to current circulation, and then describes the principal exploitation activities, such as fishing or offshore oil-well drilling, carried out in that sea or ocean. In this way a picture of the world ocean is built up. It cannot be a complete picture: every day a scrap more information is wrested from the oceans to add another piece to our partly completed jigsaw puzzle of knowledge and experience. The completion of that puzzle represents one of the greatest challenges left to man.

2 The Atlantic Ocean

The second largest and most well-studied of the great oceans, particularly important for rich fishing grounds and for oil and gas reserves.

Trade and war are two factors that more than any other provide the impetus for gathering knowledge about the seas and oceans. This is especially true of the Atlantic Ocean, where trade and war have been conducted vigorously for centuries. The first charts of the Gulf Stream were produced by Benjamin Franklin in 1787, not in a spirit of lofty academic idealism but because, as Postmaster-General of the American colonies, he realized that mail-carrying ships could speed their eastward voyage by taking advantage of the fast-flowing waters of the Stream. In World War II, Allied shipping convoys bringing food to hungry Britain played desperate hide-and-seek with German submarines across the North Atlantic. Much of the equipment, such as sonar, invented to play its part in that deadly game has since been developed and refined to aid scientists in their bid to unlock the secrets of the oceans.

As a result of trade, war, and scientific interest, more is known about the Atlantic than the Pacific, Indian, Arctic, and Southern oceans. The Atlantic is the second largest ocean, with an area (excluding marginal seas such as the Mediterranean and Caribbean) of nearly 32 million square miles. It has an elongated "S" shape, well defined by the coastlines of America in the west and Europe and Africa in the east. Its northern, southern, and south-eastern boundaries, however, are arbitrary, man-imposed limits: there is no sharp physical boundary between the Atlantic and the Arctic, Southern, or Indian oceans.

With an average depth of about 13,000 feet, the Atlantic is shallower than the Pacific and Indian oceans, chiefly because of the large areas of continental shelf in the north and along the American coast, particularly off South America. These

The sea boils as the island of Surtsey roars up—the result of a vast submarine volcanic eruption off Iceland in 1963. Iceland sits astride the northern end of the Mid-Atlantic Ridge—part of the world undersea mountain range along which new seabed is continually being formed.

ATLANTIC OCEAN
General Features

vast tracts of seabed lying under comparatively shallow water play a large part in the economic importance of the Atlantic.

Another distinguishing feature of the Atlantic is the lack of oceanic islands, with which the Pacific and Indian oceans abound. Oceanic islands are the type, usually of volcanic origin, that rise straight from the deep ocean floor, as distinct from continental islands such as the British Isles and the Falklands, which are really extensions of the continental land mass, separated from it by only a shallow sea. Oceanic islands in the Atlantic cover an area of only 193,000 square miles, and the best known include Iceland (the largest, with an area of 40,450 square miles), Jan Mayen, Bermuda, Azores, Madeira, the Canaries, Cape Verde Islands, Ascension, St. Helena, and Tristan da Cunha.

In 1963 there was a dramatic addition to the Atlantic's sparse complement of oceanic islands. Off the coast of Iceland the sea suddenly began to boil and within 24 hours an entirely new island—Surtsey—had roared from the ocean depths. Its appearance, the result of a massive seabed volcanic eruption, was seized upon by many scientists as just another link in a chain of evidence that was steadily being forged to confirm an exciting theory about the formation of the continents.

Another reason for the comparative overall shallowness of the Atlantic is the large amount of sedimentary in-filling of the basins that stretch down both sides of the ocean. These basins have been fairly well delineated by oceanographers, and, starting in the north-east and going in a clockwise direction, are named as follows: the West Europe, Iberia, Canaries, Cape Verde, Sierra Leone, Guinea, Angola, Agulhas, Atlantic-Antarctic, Argentine, Brazil, North American, Newfoundland, and Labrador basins.

These basins have been charted by a series of oceanographic expeditions that have combed the Atlantic since the late 1800's. Before this the Atlantic was thought to have a smooth floor, sloping down from the coast to the centre, and then up to the other side. The first evidence that this was far from being true came in the 1850's, when surveys were made of the ocean floor along the proposed route of the first transatlantic telegraph cable. The seabed was found to be as uneven as any landscape. Most baffling of all, a mountain rose sharply in the middle to form what was named Telegraph Plateau.

Over the next century research ships continued to find evidence of this sharp rise at various places in the middle of the Atlantic Ocean. Gradually, a pattern emerged. It took the form of a massive chain of mountains, with some peaks 15,000 feet high, running from north to south in a line that follows, uncannily it seemed, the shape of the continents on either side. It was named the Mid-Atlantic Ridge and was found to be part of the greatest mountain range on earth, the Mid-Oceanic Ridge, which extends 40,000 miles through the Arctic, Atlantic, Southern, Indian, and Pacific oceans. Running the length of the mountain chain is an enormous rift valley that, in places, is 600 feet deep and 30 miles wide.

During the century over which this piecing together of the

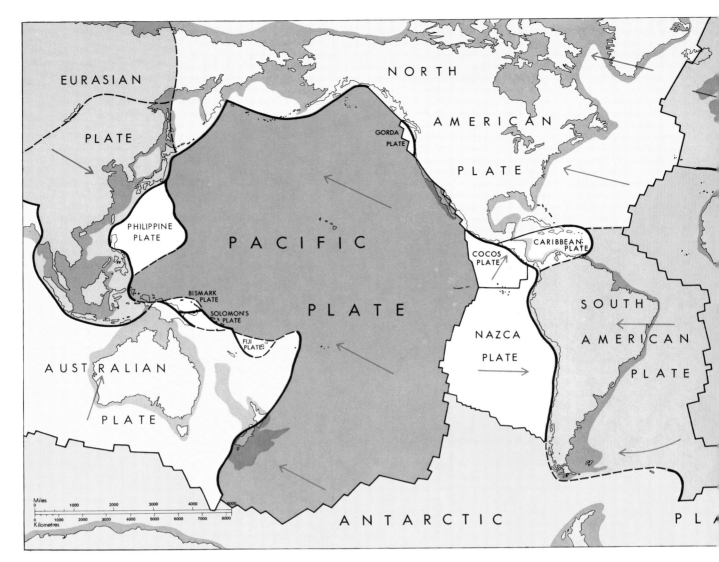

Launching the undersea probe of a magnetometer. The display console is in the foreground. The sensitive instrument measures slight variations in the earth's magnetic field, enabling the age of subocean rocks to be estimated.

world's largest, but invisible, feature was taking place, some other unusual characteristics of the ocean floor came to light. It was found that while the usual depth of the ocean basins is between 12,000 and 18,000 feet, around the margins of the oceans, notably the Pacific, great trenches occur, plunging down abruptly to depths of over 35,000 feet in places. These trenches were investigated, their features recorded, and man has been to the bottom of the deepest. However, nobody knew quite how they fitted into the overall scheme of things. But they, too, were destined to play their part in the picture that was unfolding. Further evidence came from seismologists working on the sources of earthquakes. They found that these had a world-wide distribution in which the rift valley of the Mid-Oceanic Ridge and the deep trenches were the main features.

By the mid-1960's it was plain that world scientific opinion was about to take a lurching reversal of direction, in favour of a previously scoffed-at theory—the theory of continental drift. The theory was first put forward by the German meteorologist Alfred Wegener who, in 1912, developed the idea that 200 million years ago the land areas of the earth were more or less one great island, surrounded by sea. He suggested that the continents have gradually drifted apart, moving into what is now the Pacific Ocean, and that the process is continuing. Wegener's theory, expressed in these simple terms, is now considered to be basically right. What scientists have been doing since he pro-

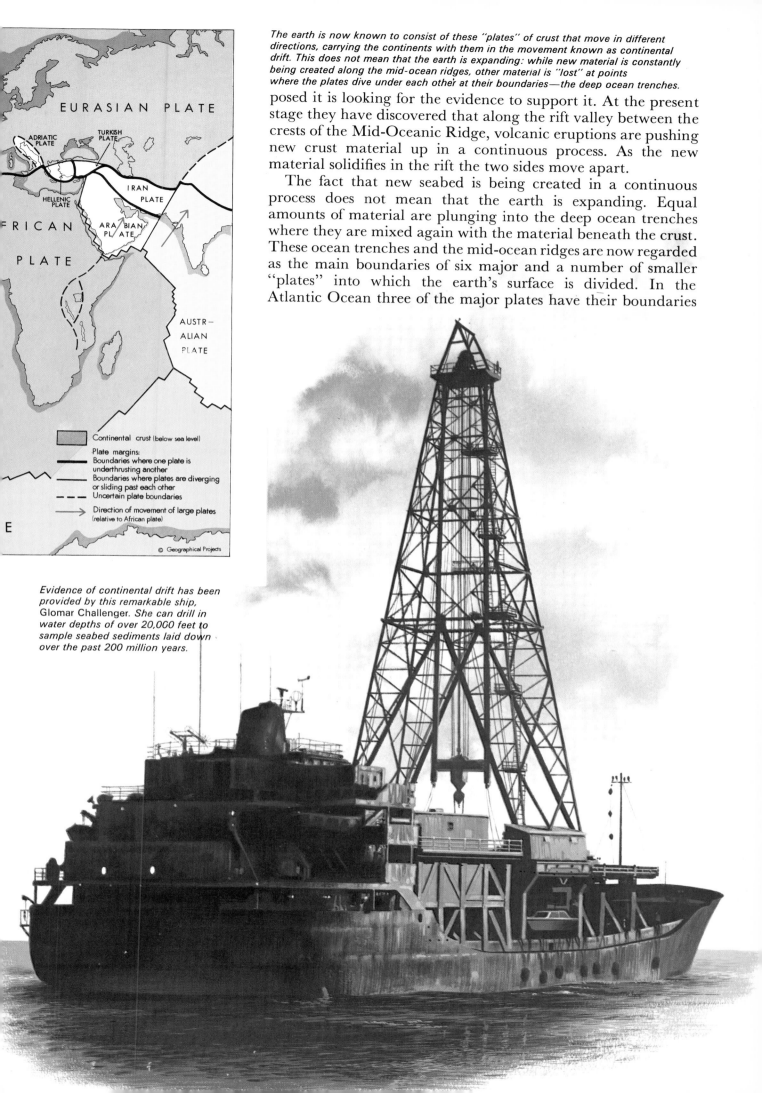

The earth is now known to consist of these "plates" of crust that move in different directions, carrying the continents with them in the movement known as continental drift. This does not mean that the earth is expanding: while new material is constantly being created along the mid-ocean ridges, other material is "lost" at points where the plates dive under each other at their boundaries—the deep ocean trenches.

posed it is looking for the evidence to support it. At the present stage they have discovered that along the rift valley between the crests of the Mid-Oceanic Ridge, volcanic eruptions are pushing new crust material up in a continuous process. As the new material solidifies in the rift the two sides move apart.

The fact that new seabed is being created in a continuous process does not mean that the earth is expanding. Equal amounts of material are plunging into the deep ocean trenches where they are mixed again with the material beneath the crust. These ocean trenches and the mid-ocean ridges are now regarded as the main boundaries of six major and a number of smaller "plates" into which the earth's surface is divided. In the Atlantic Ocean three of the major plates have their boundaries

Evidence of continental drift has been provided by this remarkable ship, Glomar Challenger. She can drill in water depths of over 20,000 feet to sample seabed sediments laid down over the past 200 million years.

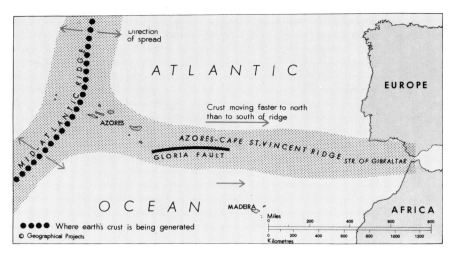

This map shows the position in the Atlantic of the 200-mile-long Gloria fault, which is over one mile wide at the top and 1,000 feet deep. It was discovered in 1971.

on the Mid-Atlantic Ridge: on the west is the American plate, to the north-east the Eurasian plate, and to the south of this the African plate. The continual movement of these plates, caused by the upsurge of new material at their trailing edges and absorption of it at their leading edges, has led to the earth being described as a group of conveyor belts, each plate representing one of the conveyors. The plate movement is very slow—it averages about one centimetre per year.

Much of the evidence for this movement is to be found only under thousands of feet of water. This is one of the reasons why the theory of continental drift has gained acceptance only in the past few years—modern technology has provided the means to test it. For example, one of the proofs that new material is being generated along the Mid-Atlantic Ridge comes from a study of the magnetism of the rocks on either side, using new and sophisticated techniques by which the age of the rock can be estimated.

These age estimations can be confirmed by taking samples of the seabed and dating the rocks in the laboratory. This is no simple task when the rocks to be sampled are covered by 3,000 feet of sediment and 20,000 feet of water! Modern technology has provided a means of doing this, however, in the shape of a specialized drilling ship called the *Glomar Challenger*, which is able to bring out cores of rock from these enormous depths. Many

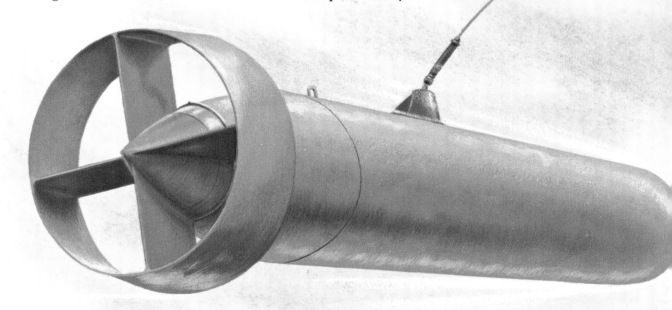

of the ideas incorporated into it are being used in commercial drilling ships as the search for offshore oil and gas extends into deeper water.

Magnetic charts and rock cores are all very well for recording the activity of the splitting sea floor, but what does the floor look like? Since the first echo sounders (or sonic depth finders) traced an outline of the Mid-Atlantic Ridge on a moving paper strip, great advances have been made in the development of "sound searchlights." Acoustic "pictures" of large areas of the seabed can now be recorded using side-scan sonar devices, which transmit sound through the water in a powerful fan-shaped beam at right angles to the direction of travel. One of the largest units of this kind in the world is operated by the UK Institute of Oceanographic Sciences. Known as GLORIA (Geological Long Range Inclined Asdic), it consists of a 32-foot-long "torpedo" which is towed at a depth of 700 feet below a ship. The sound echoes returning to the unit are recorded to give a picture of an area of the sea floor up to 12 miles long by 10 miles wide under 25,000 feet of water. Large-scale "pictures" of the sort provided by GLORIA are enabling scientists to "see" general features of the deep ocean floor for the first time, and GLORIA has already produced valuable information to support the continental drift theory.

We saw earlier that in 1787 Benjamin Franklin produced a chart of the Gulf Stream to enable ships to make faster passage across the Atlantic. His work was taken up 60 years later by another American, Lieutenant Matthew Fontaine Maury, who collected data from ships' logs to produce wind and current charts of the North Atlantic in 1847. Before his work was interrupted by the Civil War he had produced world-wide charts that were saving shipping companies up to $40 million a year in improved sailing times.

Franklin's and Maury's charts of the Gulf Stream were, as might be expected, something of an oversimplification, although they served their purpose well. Far from being a broad flowing oceanic river as they depicted it, the Gulf Stream is a series of rivers, some fast-flowing and relatively straight, others slow and meandering. The Stream begins near Florida where it is at its most powerful, pushing 40 million tons of water along at speeds of over four miles an hour. It travels north off the coast of America, and "fingers" of the current flow towards Europe and around Scandinavia, with others veering south towards Spain and Africa.

The idea of the Gulf Stream acting as a great warm river, cosseting western Europe from the rigours of an Arctic winter is also an oversimplification. True, the fingers of warm current eventually reach Norway and keep northern Scandinavian ports ice-free during winter, while ports on the other side of the Atlantic are iced up. Little is yet known about the complex interaction between air and sea and its effect on weather and day-to-day variations in the current pattern, but the generally accepted view of the Gulf Stream's effect on the climate of western Europe is that it acts as a boundary current, preventing

The 32-foot-long GLORIA sonar emits fan-shaped beams of sound through the water at right angles to the direction of travel. The returning echoes are recorded to give acoustic "pictures" of areas of seafloor up to 12 miles long by 10 miles wide under 25,000 feet of water.

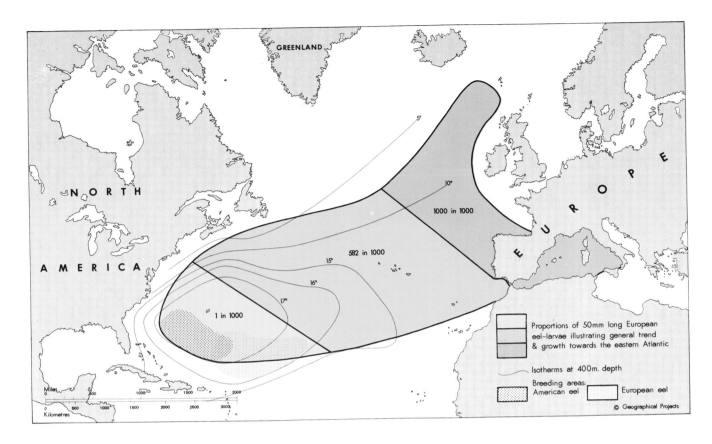

Launching a current meter. These devices are often left buoyed and unattended for several weeks to record current direction and velocity on magnetic tape for subsequent computer analysis.

the warm waters from the south overflowing over the cold water north of the Stream. South-west winds blowing over this encircled warm water keep the climate of western Europe equable.

In the centre of the huge circle described by the Gulf Stream and the other currents that flow around the Atlantic Ocean, is the warm Sargasso Sea. This area was well known to crews of sailing ships, who recognized its two main features centuries ago. In the Sargasso they feared their ships could become permanently becalmed, because there is very little wind and current movement. Their second fear was that the ships' hulls would become enmeshed in the thick weed that collects in huge amounts over an area of 1½ million square miles. This weed—*Sargassum*—is not native to the area, but is a tropical seaweed that grows near the shore of Central and South America. It breaks away and drifts into the Sargasso Sea, where, for a period, it is capable of vegetative reproduction without being attached to the seabed. Eventually it sinks and dies, to be replaced by new supplies of the drifting plants.

The Sargasso Sea also figures large in a remarkable piece of detective work undertaken in the 1920's by the Danish scientist Johannes Schmidt. Before his work, the life history of the common freshwater eel was a biological mystery. Schmidt found that mature eels migrate from European and North American streams and rivers to the Sargasso Sea, where they spawn and die. Their eggs hatch in the warm water and the larvae, which at this stage are a delicate leaf shape, slowly make their way back across thousands of miles of ocean to the rivers. The European varieties spend three years, and the American varieties one year, on this journey and on the way they gradually

Two Atlantic drifters. The common eel (above) drifts from the Sargasso Sea along the Gulf Stream to American and European rivers (see map opposite). The *Ben Franklin* (below) spent 30 days under the Atlantic drifting with the Gulf Stream from Florida to New York with six men aboard.

change into the familiar eel shape. They mature into adult eels several years after swimming up the rivers and streams.

This remarkable feature of the common eel in spending part of its life drifting in the deep ocean is not shared by many other of the fish that are caught and eaten by man. In fact, some 92 per cent of the world fish catch comes from the narrow continental shelves that make up just 10 per cent of the ocean area.

Fishing is the major resource industry of the Atlantic Ocean and the northern continental shelves support the most intense fishing activity in the world. The continental shelves of Europe, Greenland, Canada, and the eastern United States teem with fish: cod, haddock, redfish, flounder, plaice, sole, halibut, turbot, herring, mackerel—the list is long. The fish can be divided into types according to their habits. *Demersal* fish live and feed mainly on the seabed, and can be further divided into roundfish (such as cod and haddock) and flatfish (such as sole and plaice). *Pelagic* fish live and feed nearer the surface, and include herring, mackerel, and the sardine-type fishes.

In the central Atlantic area, more exotic, often highly coloured tropical fish are sought off the coast of Africa—pelagic fish such as sardine and horse mackerel; demersal types such as bream, sparids, and hake; and cephalopods—highly prized (by some!) delicacies such as squid, octopus, and cuttlefish. All these

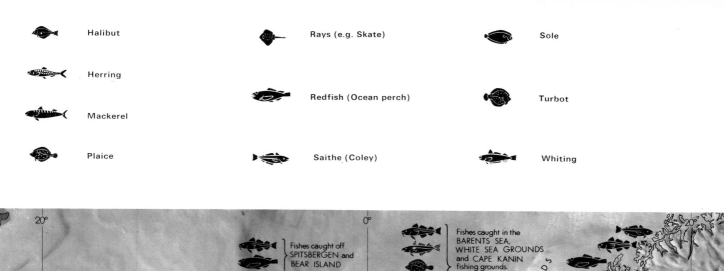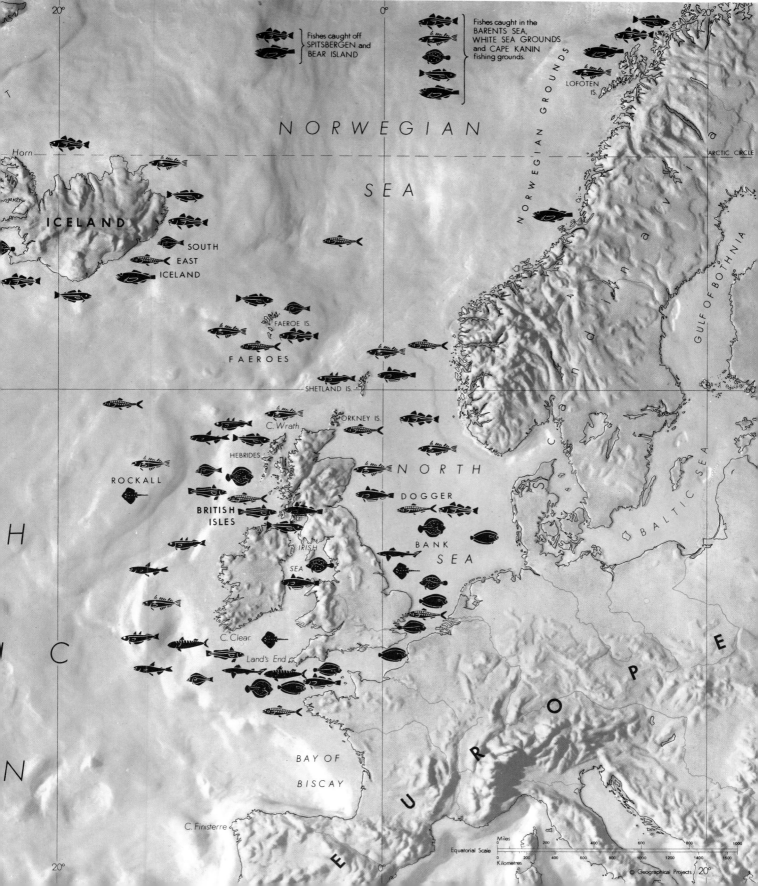

are sought not only by local fishermen in small boats but also by the far-ranging vessels of large fishing fleets from eastern European nations such as Poland, East Germany, and Russia, as well as those of Cuba and even Japan.

These larger vessels also hunt one of the few commercially caught deep-ocean pelagic fish species—tuna. Many of these foreign fleets use the Canary Islands to trans-ship their refrigerated catches into large cargo-carriers, which make regular runs from Africa to Japan and Russia. As a result, the Canaries port of Las Palmas has become a cosmopolitan town where trawler captains from Leningrad rub shoulders with their Spanish and Japanese counterparts. Today yet another accent can occasionally be heard—the Texan drawl of an oilman, on leave from his drilling rig off the coast of Africa, where exploratory drilling for oil is in full swing.

The international fleets chase tuna right across the Atlantic to northern South America, where the important shrimp fishery of the Caribbean and Gulf of Mexico dominates. Large stocks of certain demersal fish, such as croakers and snappers, are thought to exist off Surinam and Guyana, but are at present largely unexploited.

Farther south again, off Argentina, is the largest single area of continental shelf in the whole Southern Hemisphere—the Patagonian Shelf. The only visible part of the shelf is the Falkland Islands, the tiny British dependency over 300 miles from the mainland. North of the Falklands the most important species of fish is a type of hake that has increasingly attracted the attention of the world-ranging fishing fleets of Spain and eastern Europe. Ships from these nations first went south in search of the hake that abounds off South and South-West Africa: there, as a result of Spanish and Russian activity, catches of Cape hake increased from around 100,000 tons a year in 1963 to over 400,000 tons a year in 1967. Some fisheries experts think that the shelf area north of the Falkland Islands could support as big a fishing industry as that of the North Sea. Certainly, the climate and general conditions are similar, and surveys have shown that the fish species are not too dissimilar to those of north-west Europe. An example is the Falkland herring, stocks of which, it is estimated, could support an annual catch rate of over one million tons.

But it is in the north that the main onslaught is at present being made on Atlantic fish stocks. The development of North Atlantic fisheries has been governed by technology and economics. With the exception of the Portuguese, who as early as the mid-1400's were fishing the Newfoundland Grand Banks, the fishing industry has moved out to sea a step at a time. The industry had its beginnings in the North Sea. The English towns of Hull, Grimsby, and Barking, for example, were important fisheries centres in the 1100's. Over the centuries larger boats were built and fishermen began to move farther out into the North Sea. By the late 1800's a combination of steam power and more efficient trawling gear meant that the rich fishing grounds of Iceland were opened up to ships that had previously fished only the North Sea. Then came the diesel engine and the search

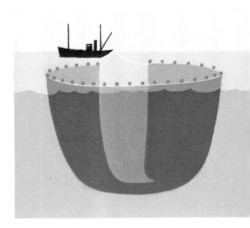

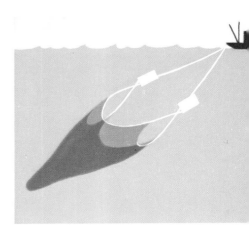

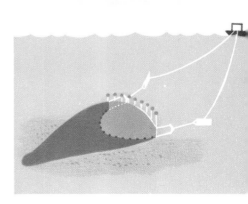

Three ways to catch a fish. In purse seining (top) an entire shoal of midwater swimming fish such as herring can be surrounded by the huge net; the midwater trawl (centre) skims through the shoal at a controlled depth; the bottom trawl scoops demersal fish such as cod and haddock from the seabed.

Right: Important North Atlantic species.

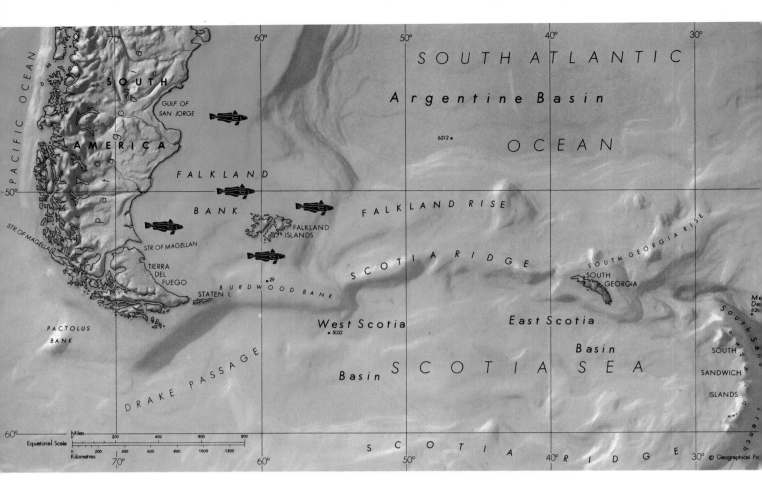

for cod extended beyond the Arctic Circle to as far east as Murmansk, in northern Russia, and as far west as Greenland.

After World War II, Germany, France, Spain, Belgium, Russia, Poland, and the United Kingdom conducted a wide-ranging and increasing onslaught on the North Atlantic grounds and the stocks of fish began to show for the first time that they were not inexhaustible. By the mid-1950's trawler skippers found that they were having to stay longer on the grounds or venture farther and farther afield in order to fill their holds. But the catch was preserved only on ice in an insulated fish hold, so there was a limit to the time they could stay on the grounds before the first-caught fish began to deteriorate.

Technology provided the answer to this "long voyage/bad fish" dilemma in the form of quick freezing on board ship. The first commercial vessel to adopt this method of preserving the catch was the British *Fairtry 1* in 1954. Another innovation in this vessel was the stern-trawling technique in which the net was hauled up a ramp at the stern (instead of over the side of the boat) and the fish emptied through a hatch to a processing deck below. This gave more room on the trawl deck for handling the gear, which meant that still larger trawls could be used, and also gave protection for the crew to gut and wash the fish. (On side-fishing trawlers fishermen have to stand for hours in freezing Arctic winds and heavy seas to haul the gear, sort, gut, and wash the fish, and mend the net.) Other nations soon followed Britain's example and today there are hundreds of freezer stern trawlers roaming the North Atlantic, using sophisticated electronic aids.

In the pelagic fisheries, too, equally great strides have been made. The traditional method of catching herring, by drift-

As the stocks of fish on traditional grounds in the Northern Hemisphere—particularly the North Atlantic—become more and more depleted, the fishing fleets of many nations are turning to the southern seas for new sources of supply. One such area is the Patagonian Shelf, a huge expanse of continental shelf off South America, where there are vast stocks of hake.

A block of frozen fish is unloaded from a plate freezer on the covered factory deck of a modern freezer stern trawler, while freshly gutted and washed fish are loaded into the adjacent unit. Quick freezing means that trawlers car stay away from port for periods of up to three months at a time with no risk of the fish spoiling.

netting, has been largely replaced by purse seining. This uses a huge curtain of netting up to three quarters of a mile long and 500 feet deep that encircles a whole shoal of fish. The net is closed at the bottom to form a huge basin and then winched in through a hydraulically powered pulley-like device known as a power block. Pelagic, or midwater, trawls have been developed, too. These are towed at varying depths, from just below the sea surface to just off the bottom. Boats using purse seines or pelagic trawls are fitted with sonar devices which send out sound pulses over distances of up to three miles. The returning echoes recorded on board the boat indicate the size and depth of shoals of pelagic fish. On midwater trawls other acoustic devices keep track of the fish up to the point where they enter the net.

Continuing technical development in a fishery can reach the point where it forms a vicious circle: as the stocks of fish decline, more and more efficient methods are required to maintain catch yields. But the more efficient the catching methods become, the more the stocks of fish decline. This stage has been reached, or is approaching fast, with such important North Atlantic fish as cod and herring. As far as these species are concerned, the trend can be reversed only by better fisheries management, including such measures as the use of larger meshes in nets, so that only mature fish of a certain minimum size are caught, and a limit on the amount of fish to be caught. This implies a degree of international control that is difficult to maintain, but such is the concern over the state of fish stocks that a catch quota system was brought into operation on certain North Atlantic grounds in 1973, and this will probably be widely extended over the next few years.

This does not mean that the total fish catch in the North Atlantic will necessarily have to drop. There are some species of fish that are either capable of far greater exploitation, or which are hardly exploited at all. This is because there is little demand for them, even though it has been shown that the consumer can be educated to try new types of fish. The Germans, for instance, are one of the few nations to eat redfish (also called ocean perch), a North Atlantic roundfish that is marketed in that country under the name of sea bream. Yet British trawler crews often throw redfish overboard when they catch them because there is little demand for them in the UK. In the USA, a type of cod known as ling cod was unpopular until somebody hit on the idea of changing just one letter of its name to make it King cod and mounting a big publicity campaign to promote its virtues. Off the west coast of Ireland there are vast stocks of two types of roundfish, the blue whiting and great silver smelt, or argentine—yet nobody hunts them in any great number.

Atlantic fisheries, then, still offer potential growth. What will probably need to be changed to keep them going is the catching emphasis and consumer tastes. The Atlantic, as we have seen, is a huge ocean. Providing its resources—whether fish, or the practically untapped oil and mineral riches—are exploited rationally and with foresight, it will continue to contribute greatly to man's prosperity.

3 The Pacific Ocean

 A rich and varied ocean, from warm coral lagoons teeming with marine life to cold abyssal plains rich in minerals, split by deep trenches.

Widest, deepest, longest, most productive—the Pacific is the greatest of the great oceans. So great that the figures about its size are too big to convey much meaning. Better to say, perhaps, that if you cut out a map of the United States of America you could fit it into a map of the Pacific about seventeen times; that if you cross the Pacific from Panama to the Gulf of Thailand you have travelled halfway around the world; that you can travel from the freezing waters of the Arctic, across the equator, and on into the ice of the Antarctic, all without a single land mass of any significant size intervening. Its average depth is greater than that of the other oceans and it contains the deepest point on the earth's surface.

The Pacific has always represented a challenge. A challenge to plunge into its profoundest depths, to cross its greatest expanses, to harvest its wealth, to tame the often-exhibited ferocity that mocks its name.

The greatest depth in the Pacific known to man has been conquered. On January 23, 1960, the bathyscaphe *Trieste*, designed by Professor Auguste Piccard, dipped beneath the waves above the Marianas Trench, a seven-mile-deep gash in the Pacific floor some 200 miles from the island of Guam. On board *Trieste* were Professor Piccard's son, Jacques, and US Navy Lieutenant Don Walsh. It took them nearly five hours to reach the bottom where *Trieste* settled silently onto a carpet of ooze. The bathyscaphe was 35,800 feet beneath the Pacific surface. Her powerful lamps probed into a darkness that had never before been broken, and picked out—a fish. It was a flatfish, something like a sole or a ray, about a foot in length.

If any justification for *Trieste*'s voyage was required (other than just being there), this was it. For it proved that even rela-

A Pacific coral reef is a living thing made up of millions of animals who are members of a class of the lower invertebrates called Coelenterata. Each animal (a polyp) forms a stony cup for itself, and links to its neighbour, building huge colonies that constitute atolls and reefs.

tively complex creatures like this vertebrate were to be found at the greatest ocean depths, thus bringing to an end a scientific argument that had fizzed spasmodically for more than a century.

The Marianas Trench is just one of a series of deep gashes in the ocean floor that practically encircle the Pacific. They form part of a major series of active fractures of the earth's crust. This great ring of instability in the earth's thin covering makes the Pacific Ocean the most seismically active area in the world. Earthquakes and undersea volcanic eruptions occur frequently.

On land, the effects of an earthquake or volcano are immediately and often horribly evident. But when an earthquake occurs on the deep-ocean floor there is often little to disturb the surface, thousands of feet above. Nothing, that is, apart from a series of small waves, such as on a smaller scale might be generated by a stone thrown into a pond. In the open ocean these long low waves are seldom more than a foot high, although they can be 200 miles long in great arcs. They travel at incredible speeds across the Pacific—sometimes up to 500 miles an hour—in groups of about six, with up to 100 miles between crests. Because of this speed and their height they pass under ships in the open ocean unnoticed. But when they approach the shore they are transformed into killers.

They are seismic sea waves, more commonly known as *tsunamis* from the Japanese, which roughly translated means "a wave in a harbour or port." Sometimes they are called tidal waves, which is incorrect—they have nothing to do with tides. Whatever they are called has made no difference to their destructive capacity over the centuries. In 1707 a tsunami ripped into Hokkaido, Japan, and killed 30,000 people; in 1946 a tsunami triggered off by an earthquake in the Aleutian Trench took just four hours to travel 2,300 miles to Hilo, Hawaii, where it killed 173 people and caused $25 million worth of damage; an earthquake in the north Pacific in 1964 sent these deadly long waves racing down the Pacific along the west coast of the USA, causing damage of perhaps over $100 million.

How can a 12-inch-high wave devastate an area normally safe from flooding? What makes it so often much more destructive than the large waves that appear during and after storms? To answer this it is necessary to understand a little of the complexity of the waves that can be seen breaking on any shore.

To the casual observer it might seem that waves are bodies of water moving across the sea surface. However, toss a rubber ball out to sea and it can be seen that it moves up and down, but does not come towards the shore at great speed. What happens is that the wave form moves forward, not the water.

The exact origin of waves has still not been satisfactorily explained. It is known that winds above a certain speed (around three miles per hour) act on the surface of the sea in such a way as to exert pressure and to "wrinkle" the surface into ripples. The wind continues to act on the surface thus presented to it until the wave increases in height. While the wind is blowing hard, the open sea presents an awesome and confused picture. But as the wind dies, the height of open-ocean waves decreases, their crests round off, and the sea surface settles down into a

This is the devastation caused by the 1964 tsunami that hit Kodiak, Alaska. Tsunamis are waves generated by earthquakes. They cross the ocean at speeds of up to 500 miles an hour, but because they are less than a foot high, they are hardly noticeable in the open sea. When they reach the shore, however, they grow to awesome heights and rip across the land, causing the sort of damage shown here.

series of long parallel undulations. This is known as swell.

As waves or swell approach a shoreline, and the orbits of their water particles change from circles to ellipses, they slow down and those behind begin to catch up with those in front. Because little energy is lost by friction, and while slowing down they have to carry as much energy as before, their height increases. Eventually they reach such a disproportionate height to length that they become unstable and fall over, or "break," on the shore.

The effects of this slowing down of wave motion and the subsequent breaking on a shoreline can be observed on any beach. Similar changes occur when a tsunami strikes. The long low waves slow down as they approach the shore, and in shallow water their enormous energy can turn them into massive waves, perhaps 100 feet high, which then crash down on the land with killing force.

Many lives have been saved by the Tsunami Warning System. Some 20 seismic stations around the Pacific keep constant watch on seismic conditions that may generate tsunamis, while 40 tide stations record the waves themselves. The data are flashed to a central base in Honolulu, and warning is given to areas where the killer waves may strike.

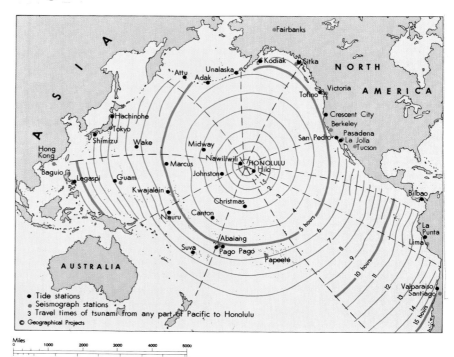

Anchovy are brailed from a purse seine that has been winched alongside a Peruvian fishing boat, aided by the skiff in the foreground. The massive Peruvian catch of anchovy goes almost entirely to make fishmeal.

Such was the destructive force of the tsunami that lashed Hawaii in 1946 that the US Coast and Geodetic Survey set up the Seismic Sea Wave Warning System in Honolulu. The life-saving value of the service was soon appreciated and today Honolulu is the centre of a network of seismographs and tide gauges that ring the Pacific. Now known as the Tsunami Warning System, it enables scientists to maintain a constant watch on the sort of seismic conditions, as recorded by 20 seismic stations, that are likely to generate a tsunami. Some 40 tide stations around the Pacific give the first signs of an actual tsunami, and recordings from these stations are flashed to Honolulu. From this information the scientists can now make accurate forecasts of the path of the waves and their likely arrival time at any point. There is still some difficulty about forecasting their heights but the prior warning that can be given has saved many lives.

Another great feature of the Pacific attributable to the instability of the ocean floor is one that figures large in the romantic history of the ocean: coral atolls.

One of the ways in which coral atolls are formed was first explained by Charles Darwin during the famous voyage of H.M.S. *Beagle* in 1831 to 1836, and his theories have been proved, by deep drilling, to be correct. The atolls are formed of coral that has gradually built up on the tops of extinct volcanoes flung up close to the surface. Coral grows in only about the top 200-foot-layer of ocean (the depth to which strong light penetrates and where the right temperature conditions for coral growth exist), yet atolls consist of thousands of feet of dead coral resting on the volcano, with a layer of living coral near the surface. Darwin theorized, correctly, that the volcano had originally been close to the surface to enable the coral to begin growing and had then gradually sunk into the earth's crust. However, the rate at which the coral grew was faster than that at which the volcano sank, and the atoll thus remains to mark the spot of a previous upheaval in the ocean floor. Changes in sea level can trigger off the same chain of events.

Atolls are not the only proof of wide-ranging and immense volcanic activity in the Pacific. Thousands of volcanoes have been flung up to heights of over 3,000 feet above the ocean floor, yet have not penetrated the surface. These are called seamounts, and there are probably about 14,000 in the Pacific, although only about one-tenth of these has yet been discovered by research ships' echo sounders and sonars. Many submerged volcanoes do not come to a sharp, mountain-like peak, but have flat, plateau-like summits, like a cone from which the top has been sliced. These summits are found at depths of between 2,000 and 6,000 feet below the Pacific surface and are known as guyots after the geologist Arnold Guyot, who discovered them. Samples dredged from the top of guyots have provided conclusive evidence that these are sunken islands, i.e. volcanoes that were thrust up above the sea surface where their tops were eroded away by winds and currents, and which have gradually sunk back into the earth's crust. A few volcanoes, however,

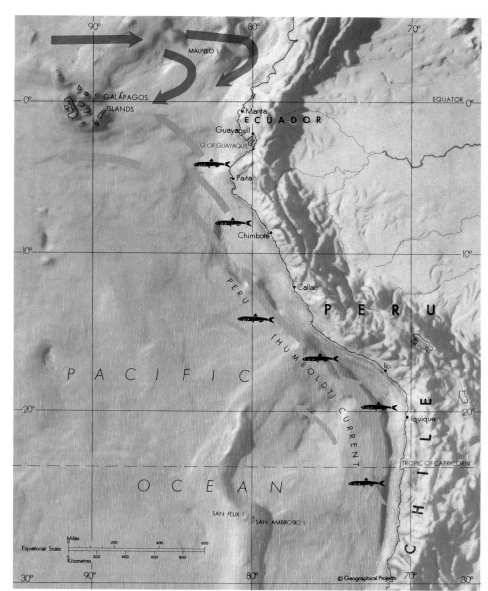

The Peruvian fishery is the biggest single-catch fishery in the world, with annual catches of around 12 million tons. This map shows the extent of the fishing activity.

have neither sunk back into the sea, nor need a cap of coral to mark their position. Hawaii is an example of a massive volcano that has climbed high above the Pacific floor. Its overall height of 32,000 feet makes it the biggest single mountain on earth. The tallest mountain *range* on earth is also in the Pacific Ocean. This is where the Andes mountains plunge from a height of 23,000 feet above the sea level down into the Peru-Chile Trench which is 25,000 feet deep, with only a narrow lip of continental shelf intervening.

The ocean along the Chilean and Peruvian coastline supports a fishery that makes Peru the greatest fishing nation in the world. Over 500 species of fish are found in the area, all supported by the northern-flowing Peru, or Humboldt, Current, and by a phenomenon known as upwelling. This occurs when the combined effects of the drag on the sea surface of the southeast trade winds, and the rotation of the earth, drive layers of surface water away from the coast. Deeper water rises to take the place of the surface water, bringing with it nutrients, such as nitrates, phosphates, and silicates, that have accumulated in the deeper water from dead organic material that has rained down from the surface layers. In the surface layers these nutrients act as fertilizers encouraging plant plankton to photo-

PACIFIC OCEAN
General Features

Three stages in the development of a coral atoll. Coral grows on the sides of an extinct volcano. The volcano gradually sinks back into the earth's crust, but the coral keeps growing. Eventually, the volcano sinks below the surface but the coral atoll remains.

synthesize and multiply. This is the first stage in a food chain that supports millions of fish.

Of the 500 fish species, just one makes Peru the greatest fishing nation in the world. This is the tiny anchoveta, a sardine-like pelagic fish that is found up to about 150 miles off the Peruvian coast. A fleet of around 2,000 boats of between 60 and 90 feet in length and equipped with every modern aid, such as sonar, power blocks, and pumps to transfer the fish from net to hold, catches 12 million tons of anchoveta every year. None of this massive catch goes directly to feed human beings: it all goes to processing factories where it is converted into fish meal for animal feed. More than half of this fish meal is exported to Europe, while most of the rest goes to North America.

There are, of course, dangers in this heavy dependence on a single species for a nation's fisheries wealth. The ecology of any part of the world ocean is always in precarious balance, and the waters off Peru are no exception. In some years, a warm, tropical current called El Niño, which usually flows south only as far as Ecuador, pushes farther south to Peru, where it tends to blanket the cold water of the Peru Current. When this southerly movement coincides with abnormally weak winds along the Peruvian coast, disaster strikes the fishing industry. The winds are not powerful enough to drive the surface layers of water out to sea, and so the cold, rich water from the ocean depths does not rise to the surface. The combination

Wherever there's a tuna there's usually a Japanese fishing boat not far behind! This map shows dramatically how the Japanese tuna longline fleet has expanded its areas of operation from 1948 until 1962, when there was not a single tuna fishery untouched by Japanese influence.

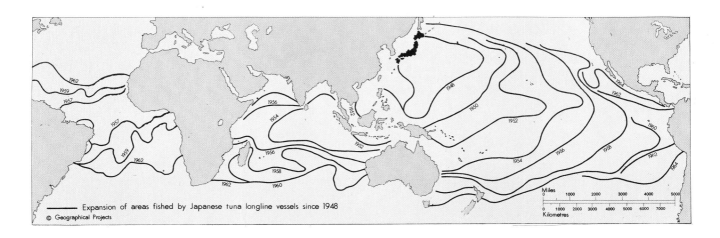

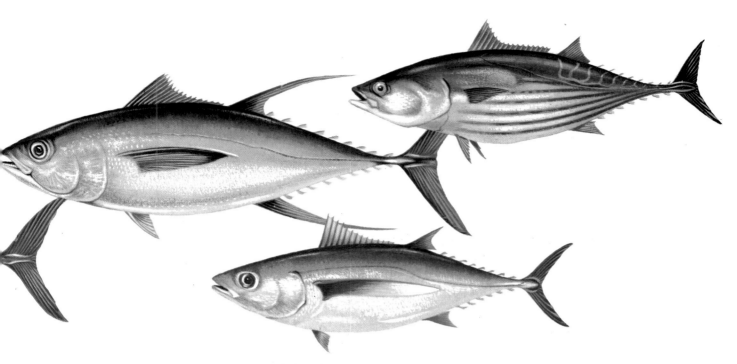

Some species of the tuna family can swim at speeds of nearly 60 miles per hour and cover 80 miles a day during migration. These beautiful fish are hunted all over the Pacific, and the important commercial species illustrated here are, from left to right, the bluefin, bigeye, yellowfin, albacore, and skipjack.

In addition to the longlining method described in the text, two other techniques are used to catch tuna. Poling involves scattering live bait near the boat to attract the fish, and then hooking them aboard with a rod. Purse seining of a type similar to that used to catch North American herring is also employed, principally by US tuna clippers working out of Californian ports. The tuna are spotted by aircraft or by a look-out in the vessel's crow's nest. Another tracking method is to look for schools of dolphins, which are nearly always associated with tuna shoals.

of El Niño's warm water and the lack of nutrients decimates the marine population: countless millions of fish die and the coastal waters are full of the stench of hydrogen sulphide gas. For the past few years the Peruvian fishing industry has gone from strength to strength; but in the summer of 1972 El Niño struck, and the fishing industry had a disastrous year.

This type of disaster will never hit the world's second greatest fishing nation, Japan. For while Peru's fishing industry is based on one species, almost everything in the sea is grist to the Japanese fisheries mill. While almost the entire Peruvian catch goes to fish meal and is exported, the Japanese catch is aimed at feeding a growing population with only a tiny area of arable land. A Japanese eats six times as much fish as an American; half Japan's supply of animal protein comes from the sea compared to about three per cent for the USA. And while the Peruvian fishery is localized, there is no major fishery in the world where Japanese vessels are not to be found.

The Japanese catch in 1970 was 9.31 million tons. Perhaps "catch" is not exactly the right word, for the Japanese harvest encompasses just about anything that is edible, or which can be processed to be made edible: from the largest animals in the world—whales—to seaweed. Japanese factory trawlers roam to South Africa for hake, to Alaska for king crab; their fleet dominates the Antarctic whaling industry; they fish for sardines in the east Atlantic, for shrimp off Surinam (South America) and for salmon off Canada. But if one fish must be chosen as representative of this far-ranging, intensive effort, it is tuna.

Tuna and other members of the scombroid family, such as marlin and swordfish, are among the most impressive ocean fish, with beautifully streamlined bodies that enable them to reach speeds of nearly 60 mph and, during migration, to travel thousands of miles at rates of up to 80 miles per day. Japan lands between a half and two-thirds of the world tuna catch, the largest proportion of this coming from the Pacific Ocean.

The Japanese are masters of the art of catching tuna by longline. This method involves casting buoyed sections of line (the main line), each about 150 feet long, which float in loops in the

water and from which hang down shorter lines with hooks at the end. Sections of main lines are joined together to form a single line that stretches up to 40 miles from the fishing boat and which can have up to 4,500 hooks on it! A single cycle of casting the line, waiting for the fish to bite, and then hauling in the line takes up to 24 hours.

The world-wide Japanese fishing effort is a good example of how that nation is making use of the sea to compensate for an acute shortage of land space. Other nations bordering the Pacific are also looking closely at the riches this ocean contains. In the United States, for example, a growing awareness of the alarming rate at which essential resources are being used up is leading to the development of marine harvesting systems that just a few years ago would have been dismissed as science fiction ideas. Take, for example, metals. World consumption of many essential metals until the end of the century will probably exceed that of the past 2,000 years. Yet lying on the floor of the Pacific are enough supplies of copper, cobalt, nickel, and manganese to satisfy the most voracious industrial appetites.

These metals occur in lumps of ore up to the size of a man's fist, known as manganese nodules because manganese is the predominant metal in them. In the 1960's a vast enthusiasm for harvesting these nodules developed. Feasibility studies for their recovery came thick and fast, and before too long the drawback became apparent. Despite the vast quantity of manganese nodules in the Pacific, the extreme depths at which they occur (10,000 to 15,000 feet) and the costs of surveying, sampling, recovering, and processing them were felt by many to present too formidable a task to be tackled for some years to come.

Nevertheless, some organizations persevered, and at the Oceanology International Conference in England in 1972, Raymond Kaufman, vice-president of Deepsea Ventures Inc.,

Deepsea Miner *is one of the first ships to be equipped for pilot-scale manganese nodule mining. The nodules (foreground) are potato-sized mineral lumps that litter the floor of the Pacific in their millions. They are first located by means of underwater TV, the pictures monitored on board ship (top left), and the information used to direct a giant "vacuum cleaner" device, which sucks up the nodules.*

USA (part of the huge Tenneco Oil Company), was able confidently to forecast that by the end of the 1970's at least two or three commercial manganese nodule mining operations would be under way.

Kaufman was careful to spell out the difficulties facing the nodule miner; nevertheless, following trials off Florida, in comparatively shallow water over the Blake Plateau, his company is going ahead with a nodule mining program. Other industrial organizations with similar plans include such giants as the Lockheed Aircraft Corporation, the Hughes Tool Company, and the International Nickel Company. A large West German group began exploration work in 1972, and a Japanese consortium plans to mine 500 tons of nodules per day from the Pacific. The early proponents of nodule mining may have had their predictions and hopes trimmed a little by commercial survey results, but with the present level of serious interest in manganese nodules they have good reason to be satisfied.

Several companies are also showing interest in the sediments that cover the deep-ocean floor of the Pacific and other great oceans. Near the shore, on the shallow continental shelf, deposits of sediment are made up of material that has been washed from the land. In the deep ocean, deposits are mainly made up of material that has rained down on the bottom over millions of years.

These deep-sea sediments are thousands of feet thick in places, and by sampling them with a coring device the ocean scientist can learn much about the formation of the ocean floor, the types of life that existed in it thousands of years ago, the climate at the time, and major shifts of the continents.

The names of the main types of sediment are taken from the tiny animal and plant plankton whose skeletal remains form their main ingredients. These sediments have many possible uses that might make them worth recovering. Calcareous oozes, for example, could be used in the manufacture of cement; diatomaceous ooze in concrete manufacture, as an insulating material, and as an abrasive; while of the non-biological sediments, red clay may one day be worth investigating as a source of metals. Nobody has any immediate plans to tap these deep-ocean sediments, but 15 years ago nobody would ever have envisaged harvesting manganese nodules from 12,000 feet below the Pacific surface.

We saw in Chapter 2 and earlier in this chapter how the principal earthquake and volcanic belts of the earth are closely linked to the ocean trench and mid-oceanic ridge system. In the east Pacific, the term "mid-ocean" is a misnomer, for the ridge runs straight into the land at California at a point where the American and Pacific plates join. The ridge used to be in the mid-ocean, but America has slowly moved towards it as the Pacific floor has been destroyed in the process of continental drift. The American and Pacific plates rub uneasily against each other at the San Andreas fault, which is the cause of the Californian earthquakes that threaten San Francisco with destruction.

PACIFIC OCEAN
Sediments

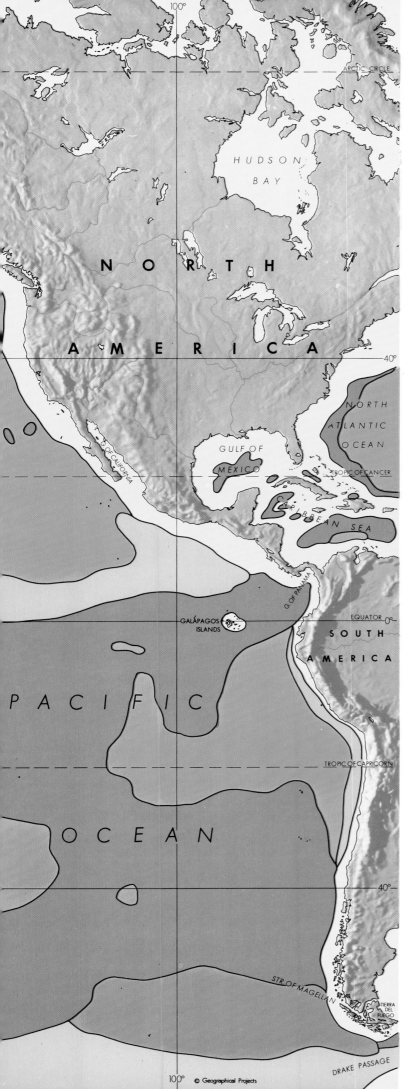

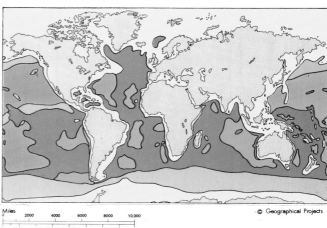

- Terrigenous deposits
- Red clay
- Calcareous oozes { Globigerina, Pteropod, Coccolith }
- Siliceous oozes { Radiolarian, Diatomaceous }

Deep-ocean sediments in the Pacific. Terrigenous deposits are those that have been washed from the land. Red clay is a deposit with less than 30 per cent organic remains. Deposits with more than 30 per cent organic remains are classified as oozes. Calcareous oozes, which are found mainly at depths of 6,000 to 12,000 feet, are made up principally of calcium carbonate. Siliceous oozes, found mainly at 12,000 feet to 15,000 feet, are composed mostly of silica.

Perhaps because of this seismic knife-edge on which they live, the Californians are very conscious of the effects of tampering with the natural environment. Their fears were confirmed in 1969 when an oilwell drilled in the Santa Barbara Channel blew out and hundreds of thousands of gallons of crude oil spewed into the sea and onto beaches. The resulting mess cost millions of dollars to clean up and brought the environmentalists and the oilmen into direct conflict. The conflict demonstrates neatly the dilemma we face in the increased exploitation of the earth. On the one hand the oilmen argue cogently that if sufficient quantities of oil and gas are not found in and around the United States, the American public will have to pay twice as much for energy requirements within a very short time. (It has been estimated that the USA faces a shortage of 58 per cent of its total energy needs by as early as 1985.) Faced with that kind of crisis, they say, the very quality of life that the environmentalists seek to preserve is threatened. The environmentalists, on the other hand, say that no price is too high to pay to prevent another accident on the scale of the Santa Barbara blowout, or to prevent unsightly drilling rigs springing up off the coast, feeding ugly refineries and processing plants ashore.

As exploitation of the Pacific Ocean and all the world oceans increases, conflicts like these are bound to arise. They will be resolved only by carefully balancing ever increasing human needs against the amount of plundering the earth can tolerate.

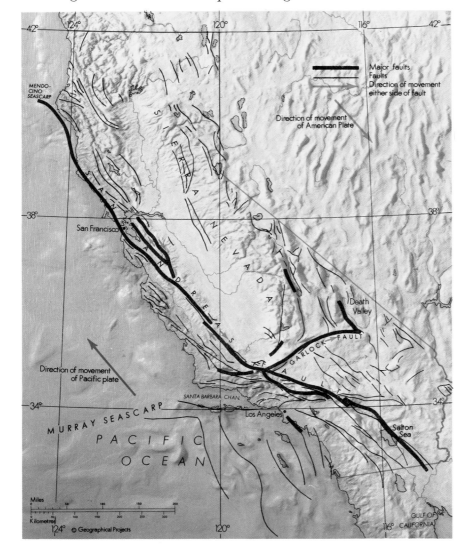

In the San Francisco-Santa Barbara area of California, the American and Pacific continental plates rub uneasily against each other to form the San Andreas fault—a geologically unstable area that threatens San Francisco with an apocalyptic earthquake.

Straw is being loaded onto dinghies to be spread on oil coming from a massive leak in the Santa Barbara Channel in 1969, in a desperate attempt to prevent the beaches of California being further despoiled. It was hoped that the straw and oil would form a heavy mixture that could either be scooped up, or would sink. The accident has come to typify the dilemma we face in maintaining both our standard of life and the standard of our environment.

4 The Indian Ocean

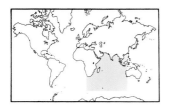

An ocean dominated by monsoon winds, which change the pattern of surface currents during the year. Recent study has revealed untapped reserves of fish.

The Indian Ocean is usually referred to as the smallest of the three great oceans. But "smallest," of course, does not necessarily mean "small." Along a line from Durban in South Africa to Perth in Western Australia, the Indian Ocean is over 4,000 miles wide. And, with an area of about 29 million square miles, it is about eight times as big as the United States.

Only in comparatively recent years has the Indian Ocean been explored to any extent (the southernmost waters were first crossed only in 1772, by Captain Cook), so that our knowledge of its physical characteristics is not nearly so well advanced as for the Atlantic and Pacific.

The first major attempt to win extensive data from the Indian Ocean was the famous world-ranging H.M.S. *Challenger* expedition of 1872–76 almost a century after Benjamin Franklin had made his charts of the Atlantic Gulf Stream.

The biggest and most concentrated co-ordinated effort to gather knowledge took the form of the International Indian Ocean Expedition of 1960–65, when the pride of many nations' oceanographic research fleets converged on the ocean. Included in the expedition were many famous oceanographic ships, such as the *Vema, Conrad, Argo, Te Vega, Anton Bruun, Discovery, Challenger II, Commandant Charcot*, the *Lomonosov*, and the *Vityaz*.

The picture of the Indian Ocean is, therefore, still being pieced together. What *is* known is that the continental shelves are generally quite narrow when compared with those of the Atlantic, except in some areas, such as off Bombay, where it reaches a width of over 125 miles. The deep ocean floor (average depth is 12,700 feet, reaching a maximum of just over 24,400 feet) has the one great feature common to the three great oceans: the towering Mid-Oceanic Ridge, which sweeps down

There is tragic irony in the fact that the Indian Ocean, teeming with virtually unexploited fish stocks, should lap the shores of nations where protein deficiency is rife. Yet fishing methods are still primitive: in the background, fishermen handle a beach seine net, used to encircle fish shoals near the shore. In the foreground, nets are mended on a dugout canoe—the one concession to modernity being the outboard motor supplied by a United Nations agency.

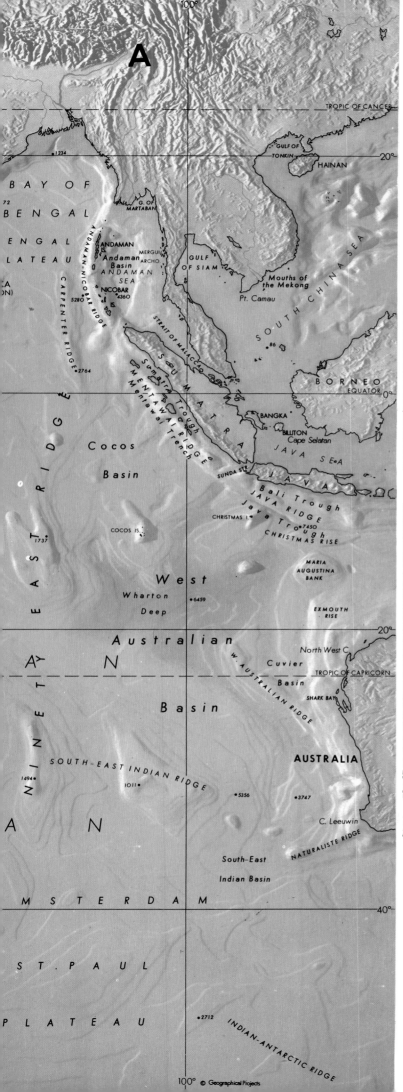

INDIAN OCEAN
General Features

Surface Currents/February

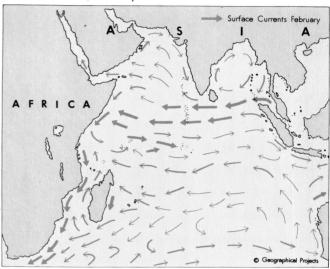

Surface Currents/August

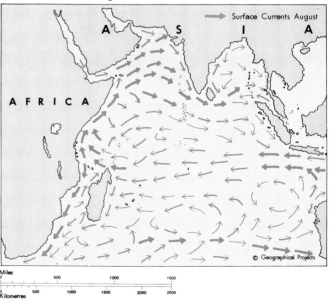

The surface current pattern of the Indian Ocean in winter is reversed in summer by the effect of the monsoons.

the Atlantic, around Africa, and into the Indian Ocean. Here, its shape is less well defined than in the Atlantic and Pacific, but broadly it takes the form of an inverted "Y". The shape of the ridge is broken by huge fracture zones, so that instead of the relatively smooth line of the Mid-Atlantic Ridge, in the Indian Ocean there appears to be a series of ridges, many of which have been given individual names such as the Carlsberg Ridge and the Ceylon Ridge.

The shape of the Mid-Oceanic Ridge in the Indian Ocean is, however, made still more confused by another notable feature. For not all the Indian Ocean ridges are part of the Mid-Oceanic Ridge system—there are a large number of "aseismic ridges." These run, generally, in a north-south direction and lack the seismically active rift valley of the Mid-Oceanic Ridge. One such ridge, only discovered during the International Indian Ocean Expedition, is named the Ninetyeast Ridge because of its longitudinal position. It runs south in a straight line from the Bay of Bengal (where its northernmost end had been discovered previously and named Carpenter Ridge) to around 32°s—a distance of over 1,500 miles. This has earned it the distinction of being called "the longest rectilinear geo-tectonic structure in the world," which means, simply, that nobody has yet found a feature on the surface of the earth which is as straight and longer.

Of particular interest and importance in the Indian Ocean is the circulation of currents and their relation to the climate of Asia. To appreciate the peculiar pattern of water movement in the Indian Ocean, however, we must first look at the general pattern of current circulation across the earth's surface.

Currents are produced by winds and density differences between water masses. Winds are caused by the replacement of hot air rising at the equator by cool air flowing from north and south towards the equator. Because the earth is rotating, a phenomenon known as the Coriolis Effect twists the air flows, so that in the Northern Hemisphere the winds are deflected to blow from the north-east towards the south-west, while south of the equator they blow from the south-east to the north-west. These are the north-east and south-east trade winds, and they push water before them to converge at the equator. The water, too, is subject to the Coriolis Effect.

Just north and south of the equator, then, two east-to-west flowing currents are set up, and in each of the three great oceans these are known as the North Equatorial and South Equatorial currents. In all three oceans this westward movement of water is arrested by a land mass—America in the Atlantic, Asia in the Pacific, and Africa in the Indian. This causes what is literally a piling up of water on the eastern coasts of these continents, and the water "overflows" to north and south. It is kept going northwards by the pressure and density gradients parallel to the coast. In the region of westerly winds they turn to the east and then to the south in a great clockwise circulation, or "gyre."

In the Northern Hemisphere there are two great gyres in

"Aimed trawling" is one technique that could vastly improve the efficiency of pelagic fishing in the Indian Ocean. Here, the boat's forward-looking sonar under the bow first spots the shoal. The downward-looking sonar establishes the shoal's depth as it passes beneath the boat. Forward-looking sonar on the net itself "watches" the fish approach, while the upward-and downward-looking sonar on the net is used so that the vessel's captain can know its depth in the water at all times and "watch" the fish going into the net.

Two species of fish caught in the Indian Ocean, both of which grow to a length of about eight inches. Top: The oil sardine, Sardinella longiceps. Bottom: A member of the mackerel group of fishes, Rastrelliger kanagurta.

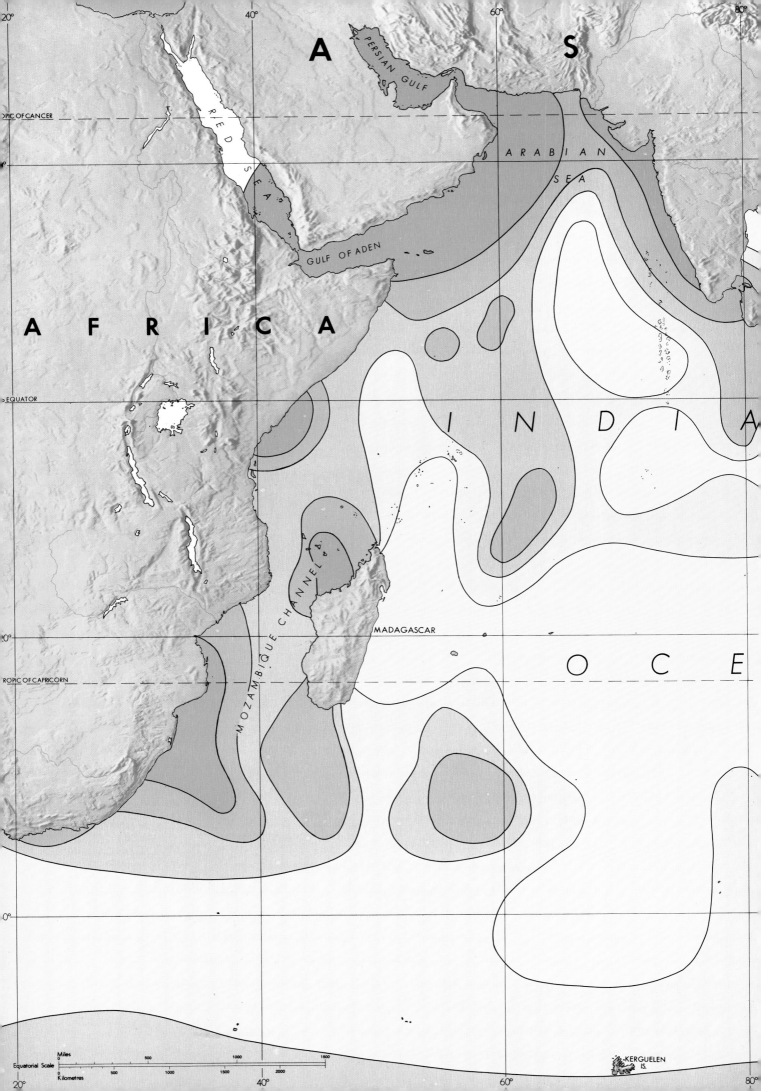

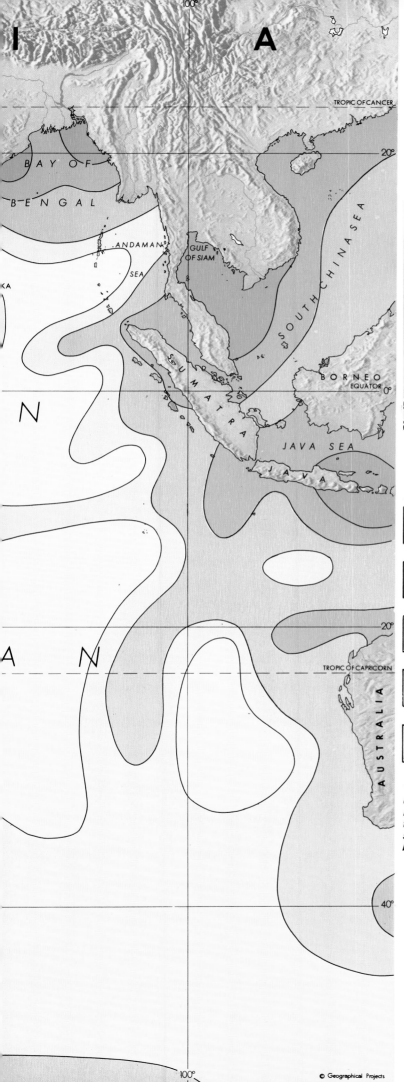

INDIAN OCEAN
Productivity

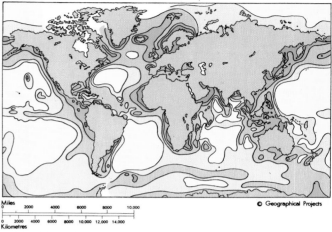

Phytoplankton primary production in milligrams of carbon per square metre per day

- More than 500
- 250 to 500
- 150 to 250
- 100 to 150
- Below 100

The fertility of the Indian Ocean (left), and world oceans (above), expressed in terms of phytoplankton primary production. Research into primary production is particularly important to those nations bordering the Indian Ocean, where protein is scarce, because it provides a pointer to areas where there may be untapped sources of fish.

PERSIAN GULF
Oil

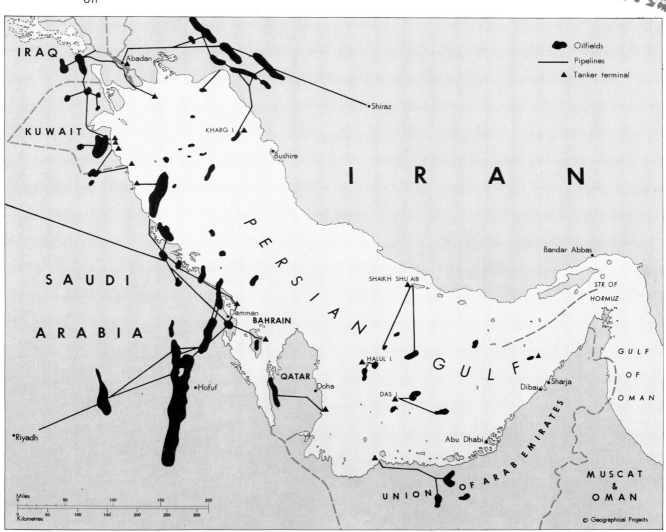

Spotter aircraft are already used to locate tuna shoals and to guide fishing boats to them. In the future, this information could also be gathered by satellites, together with oceanographic data. The data would be relayed to a shore-based computer, as would data gathered by automatic buoys, and the computed results relayed to fishing boats in the form of "best-chance" fishing areas. Satellites and ocean data buoys will also play a large part in providing data for better weather forecasting, optimum ship routeing, and in giving ocean scientists far more information about long-term physical, chemical, and biological oceanic processes.

The Persian Gulf area, showing oilwells and pipelines. Some of the most technically advanced feats of marine engineering have been undertaken in this area to tap underground oil reserves.

which the currents travelling in a clockwise direction are strongest on the western side of the oceans. These are the Kuroshio, or Japan, Current in the Pacific, and the Gulf Stream in the Atlantic, both warm currents. In the Southern Hemisphere, counterclockwise gyres are set up, the strong, warm currents flowing down the western sides of the oceans. In the Indian Ocean this southern-flowing current is known as the Agulhas Current. It flows on either side of the Island of Madagascar to about 40°s, where it curves away from Africa. Off Durban, where it is strongest, it is about 62 miles wide and, according to the most recent estimates, transports about the same volume of water as the Gulf Stream.

Part of the Indian Ocean is north of the equator, and so it would be reasonable to expect a year-round clockwise gyre similar to that of the North Atlantic and North Pacific. But a natural conflict of huge proportions rages across this section of the ocean, brought about largely by the presence of the Asiatic land mass to the north. The whole oceanic pattern north of about 2°N in the Indian Ocean is governed by the monsoons. From April to October, the Asiatic land mass heats up quickly, and the resulting low air-pressure draws in winds from the south-west. During this South-west Monsoon, the main current in the northern Indian Ocean is the eastward-flowing South-west Monsoon Drift.

During the northern winter a dramatic reversal occurs. The land cools down at a far greater speed than the ocean so that the air above the land becomes cooler and denser. Air from this high-pressure area rushes to fill the "vacuum" created by the warm, low-pressure air above the sea. On their way these Northeast Monsoon winds push water away from the land. This water is deflected by the Coriolis Effect to the west to form the North Equatorial Current (or North-east Monsoon Drift), which piles up against Africa to produce the weaker, eastward-flowing Equatorial Counter-Current.

This dramatic reversal of these mighty oceanic drifts was one of the aspects of the Indian Ocean studied by scientists during the International Indian Ocean Expedition of 1960–65. The results of the expedition confirmed many previous discoveries that hold out great promise for the future of the underdeveloped nations bordering the Indian Ocean. We have seen, for example, that the principal factor governing the huge Peruvian anchoveta fishery is the upwelling of nutrient-rich cold water along the western South American coast. Off the coast of Somaliland, particularly in the Gulf of Aden, and also off Arabia, very strong upwelling has been observed during the South-west Monsoon. This indicates that there could be vast stocks of pelagic fish near the coast in these areas. What is now needed is a major research program to discover their extent, and to provide the boats and trained crews to exploit them. It is sad that the Indian Ocean, bordered by nations where protein deficiency is often a common occurrence, should contribute less than five per cent of the world's total fish catch.

Valiant efforts have been made by such international bodies as the Food and Agriculture Organization of the United

Nations (FAO) to help fishermen, particularly of the Indian sub-continent, to improve their lot by providing them with better catching and processing facilities. But this assistance is often limited to the supply of outboard motors to power dugout canoes. Even though an effort as small as this has shown encouraging results, it is not enough. Sidney Holt, formerly director of the Division of Fishery Resources and Exploitation in FAO, feels that the Indian Ocean might become the part of the world where advanced pelagic fishing techniques are first used on a large scale. "Certainly there is evidence," he says, "of a great but scarcely utilized pelagic resource in that ocean, and around its edge are human populations sorely in need of protein."

The sort of techniques Holt has in mind would make use of pelagic trawls, possibly fishing at far greater depths than existing nets—perhaps down to 1,000 feet. These nets would use an existing European technique of "aimed" trawling. With this method, the skipper of the catcher boat searches for a shoal of fish in the vicinity and steers his boat towards it until the shoal is beneath him and can be spotted with the echo sounder in the boat. From the information supplied by the echo sounder he can judge the depth of the fish below the surface (or their height above the seabed) and so adjust the depth at which the trawl net is set. On the net there is a three-way echo sounder, which pings out sound pulses in three directions—up, down, and forward, and from the information gained from echoes produced by this the skipper can "aim" his trawl at a shoal of fish with a fine degree of precision.

This technique could be backed up by remote tracking devices that would locate the shoals of fish more quickly. Already satellites are being used to plot ocean temperatures and current movements, and there are plans to set up a network of ocean buoys that would make oceanographic measurements and send the information by radio and satellites, which would in turn relay it to receiving stations on the shore. The shore stations would sort out the information and then re-transmit it to the fishing fleet. As well as relaying messages in this way, satellites could be used directly to spot shoals of pelagic fish in the same way as aircraft are now used to spot tuna.

The use of satellites, information-collecting buoys, and shore-based stations equipped with computers to direct fishing operations hundreds of miles from land may seem a far cry from the dugout canoe with an outboard motor working a short distance off the beach, and certainly there can be no fast transition from these primitive methods to space-age technology. However, a steady program of financial and technical assistance from the developed fishing nations to the developing ones could see these methods brought into operation, perhaps in less than a decade.

Some of the most advanced engineering feats in the world are to be found close to traces of some of the earliest civilizations. These are in the Persian Gulf, one of the shallow arms of the Indian Ocean. At the north-western corner of the Gulf,

These huge tanks sit on the seabed off the Sheikdom of Dubai. Each is 205 feet high and 270 feet in diameter. They are used to store oil pumped through pipelines from the nearby Fateh oilfield until it can be offloaded by supertankers to be transported across the world. The tanks are bottomless, and as the tankers take the oil out, water enters through holes around the base; as oil is pumped in at the top, water escapes through the base. When each tank is full the oil flow stops automatically.

A diver "flies" away from an underwater oilfield at Zakum, using a powered sled. The oil is produced automatically and the performance of the system is monitored by a shore-based office 70 miles away. Divers occasionally go to check the system.

between the Tigris and Euphrates rivers, the fertility of the irrigated land led to the rapid rise of the Sumerian civilization, which was fully established in 3500 B.C. Thousands of feet below the waters of the Persian Gulf and the land surrounding the Gulf a different type of fertility—the richness of oil-bearing rocks—now makes a major contribution to maintaining our present civilization.

Over two million barrels of oil a day bubble up from undersea oilfields in the Persian Gulf, making that relatively small area of ocean the biggest offshore producer of oil after the United States and South America. And as in the past local conditions spurred the Sumerians to develop advanced aids to support their exploitation of natural resources, so today conditions in the Gulf have enabled innovatory techniques and equipment to be developed both to exploit undersea oil and to provide means of exporting land-based deposits.

The Persian Gulf is small for the amount of activity it supports. Its area is 92,500 square miles, it is 615 miles long and it varies in width from 210 miles to 35 miles. It is in many ways an almost ideal area in which to experiment with new ideas in marine engineering. First, it is shallow. Average depth is 84 feet and nowhere is it deeper than about 550 feet. Rainfall is low, fogs are rare, and the strongest *regular* wind, the *shamal*, which blows from the north-west, very seldom reaches gale force. Tides vary through the Gulf, but the greatest range is 10 feet; currents are generally of low velocity and waves are small. On the debit side, the temperature often soars well over the 100°F mark, relative humidity is uncomfortably high, and the autumn squalls can whip up waterspouts accompanied by winds that can scream up to 95 miles per hour in as little as five minutes.

These disadvantages are considered a small price to pay for the wealth that the Persian Gulf offers, and the area bristles with the often strange shapes of modern technology. For example, at the Fateh oilfield, 60 miles off the tiny Sheikdom of Dubai, three huge bottomless tanks, each 205 feet high and 270 feet in diameter, sit on the seabed in 158 feet of water. Oil from the subsea field is pumped into them through underwater pipelines, and supertankers load from towers at the top of the tanks. The tanks work on the principle that oil is lighter than water: as the tanker takes the oil out, water enters through holes round the base; as oil is pumped in at the top, water escapes through the base. When each tank is full the oil flow stops automatically.

Elsewhere, huge tankers act as storage barges for subsea oil, pumping their cargo through a single buoy mooring to smaller vessels, which moor to the buoy by the bow only so that they are free to swing according to the dictates of tide and wind. There are more conventional moorings, too: at Kharg Island, off Iran, a man-made island served by underwater pipelines from the shore has been constructed so that oil from land-based wells can be loaded directly onto some of the world's biggest tankers. At Zakum, 65 miles from Das Island, the subsea oil is not brought to the surface at all, but is produced by an auto-

RED SEA
General Features

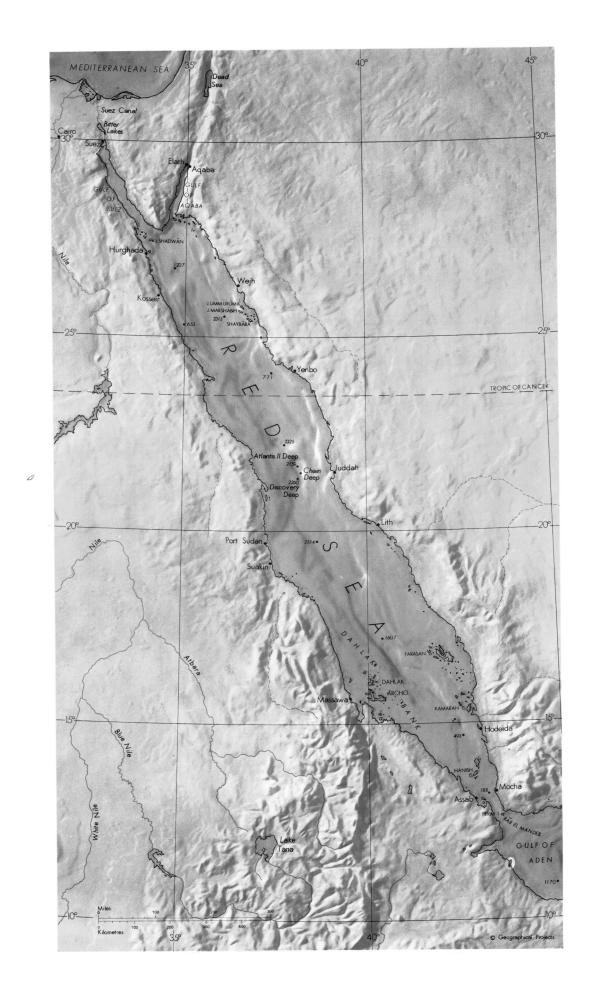

matic system on the seabed and pumped ashore through subsea pipelines. The installation lies beneath 70 feet of water and is serviced by divers, who often use torpedo-shaped underwater sleds to take them round the underwater processing plant. The performance of the whole system is monitored, through a multi-channel telemetry link, in a shore-based office 70 miles away.

Oil is also produced in another arm of the Indian Ocean, the Red Sea. But here, growing importance is being attached to a phenomenon that has been periodically examined by oceanographers over the past 30 years.

A glance at the map on page 20 shows that the boundary of two of the moving plates of the earth's crust occurs in the Red Sea. In fact, the Red Sea is a "new" sea, being created as Africa and Arabia move away from each other. As we have seen in previous chapters, this movement of plates usually means volcanic or seismic activity, and this is to be found at certain points in the Red Sea.

The Red Sea is only 191 miles across at its widest point. Yet its sides slope steeply to 1,640-foot-deep shelves which flank a narrow trough in the centre where depths of over 7,000 feet are found. It is here that the volcanic activity takes place. Scientists working on research ships have found, to their surprise, that the seabed temperatures at these great depths, far from being close to the winter temperature of the surface water (72°F), as they expected, ranged up to 133°F. When they examined samples from the bottoms of these deeps—of which the Atlantis Deep is the best known—they found them to be full of a hot salty brine, rich in iron, zinc, copper, and manganese. Beneath these brines, they discovered sediments of several hundred feet thickness which contain extraordinarily high concentrations of zinc (five per cent) and copper (one per cent), as well as other elements, including silver.

The Red Sea "hot spots," as they are known, are thus valued not only for their scientific interest but as a future source of minerals. It has been calculated, for instance, that the value of the metals in the sediments of the Atlantis Deep to a thickness of just 30 feet is about $1,500 million! And the total thickness of the sediments is about 300 feet! Needless to say, the proposals put forward to harvest this wealth have filled enough paper to soak up the entire water content of the Red Sea and get the sediments out by bulldozer. But one of these proposals will doubtless get under way soon, and out of the depths of the Red Sea a new supply of essential minerals will be won.

The Red Sea is of great interest to marine scientists because at the bottom of some of the great deeps there is a hot, salty brine, rich in iron, zinc, copper, and manganese.

5 The Arctic Ocean

The smallest ocean, now being explored for natural resources as ways are devised of coping with the ice and severe weather conditions that cocoon its riches.

Unfortunate holidaymakers, shivering on the beaches of England during the early summer of 1972, joked wryly about the coming of a new ice age—and they were not far wrong. Their discomfort stemmed from the presence in the North Atlantic of more icebergs and pack ice for the time of year since 1912. Some 400 icebergs were counted in a relatively small area of ocean off North America—four times the usual number. The cold water bringing the ice from the Arctic cooled prevailing westerly winds, and so rain and cold came to the United Kingdom. Thus, in midsummer the Arctic reminded England that it is always capable of giving the world to the south of it a chilly jolt.

The Arctic is the smallest of the world oceans and less is known about it than all the others. In fact it was discovered to be a true ocean only about 80 years ago. The reasons for this lack of knowledge are not difficult to appreciate: the area is in darkness for nearly half the year; in air temperatures of about 4°F and winds of Force 6, exposed flesh freezes; put your ungloved hand momentarily on a steel surface and the skin pulls off when you try to remove it; a trawler nosing into the pack ice in search of cod can be overtaken rapidly by gales that pile up enormous thicknesses of ice on rigging and aerials so that a ship could become top-heavy and topple over; a slight shift in wind can make a calm sea freeze, the ice trapping and crushing ships; icebergs the size of islands move silently through the darkness, twisting and lurching treacherously in the shifting currents.

Three voyages of discovery have had a tremendous impact on the way the Arctic may be exploited in the future. The first voyage is probably the best known. It was made by the Norwe-

Ice begins to form dangerously on the rigging of a trawler in the freezing Arctic night. These are typical winter conditions in the Arctic, which is the smallest of the world oceans. But it has a fascination for explorers, who attempt to sail across it—or under it.

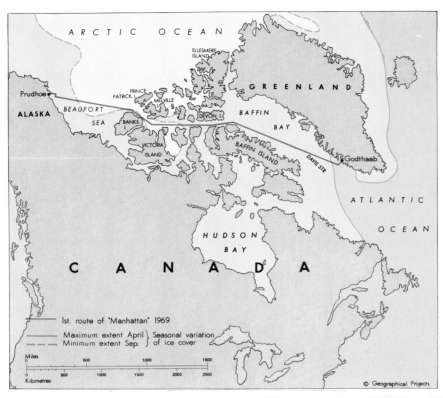

The United States tanker Manhattan, *made a 10,000-mile return journey through the Northwest Passage in 1969 (see map). She was testing the feasibility of transporting oil from the discoveries on Alaska's North Slope to North American markets.*

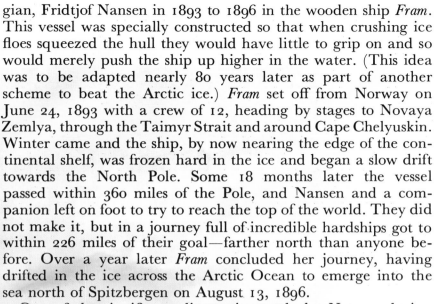

gian, Fridtjof Nansen in 1893 to 1896 in the wooden ship *Fram*. This vessel was specially constructed so that when crushing ice floes squeezed the hull they would have little to grip on and so would merely push the ship up higher in the water. (This idea was to be adapted nearly 80 years later as part of another scheme to beat the Arctic ice.) *Fram* set off from Norway on June 24, 1893 with a crew of 12, heading by stages to Novaya Zemlya, through the Taimyr Strait and around Cape Chelyuskin. Winter came and the ship, by now nearing the edge of the continental shelf, was frozen hard in the ice and began a slow drift towards the North Pole. Some 18 months later the vessel passed within 360 miles of the Pole, and Nansen and a companion left on foot to try to reach the top of the world. They did not make it, but in a journey full of incredible hardships got to within 226 miles of their goal—farther north than anyone before. Over a year later *Fram* concluded her journey, having drifted in the ice across the Arctic Ocean to emerge into the sea north of Spitzbergen on August 13, 1896.

One of the significant discoveries made by Nansen during *Fram*'s drift was the deep basin under what until then had been thought of as the Polar Sea. This basin is as deep as 13,000 feet in places, and the soundings that Nansen made from *Fram* mark the true discovery of the Arctic "Ocean." Later surveys showed the ocean to be divided by three large submarine ridges, one of which is an extension of the Mid-Atlantic Ridge. Later surveys showed the Arctic land mass and the huge continental shelf (one third of the Arctic Ocean's area) to be comprised largely of sedimentary rock. This combination of sedimentary rock and a large continental shelf has proved a magnet to the offshore oil industry, spurred on by large discoveries on adjacent land.

In order even to contemplate exploratory drilling in the hostile Arctic, oil companies have had to make careful studies of all aspects of the geology, oceanography, and meteorological conditions of the region. These studies have, in only a few years, greatly advanced our knowledge of the Arctic area as a whole, as well as emphasizing still further the formidable difficulties of working in the area.

For a start, the oil companies have found that there is nothing that could be termed a "typical" Arctic environment. Areas outside the Arctic Circle can have what might be thought "typical" Arctic conditions, with pack ice, icebergs, and fog posing a continual threat, while areas in the Arctic Ocean itself can be relatively mild and almost comfortable to work in at certain times of the year. The main areas in which either geophysical prospecting or exploratory drilling have been carried out are in the Davis Strait, in the Chukchi Sea (between Alaska and Siberia), off Alaska's North Slope in the Beaufort Sea, and around Canadian Arctic islands such as Baffin Island. The USSR also plans experimental production of oil off Siberia.

One of the most comprehensive experimental drilling programs inside the Arctic Circle was carried out during 1970 and 1971 by the drilling ship *Typhoon*. The ship worked in the Davis Strait, where one of the principal problems it en-

ARCTIC OCEAN
General Features

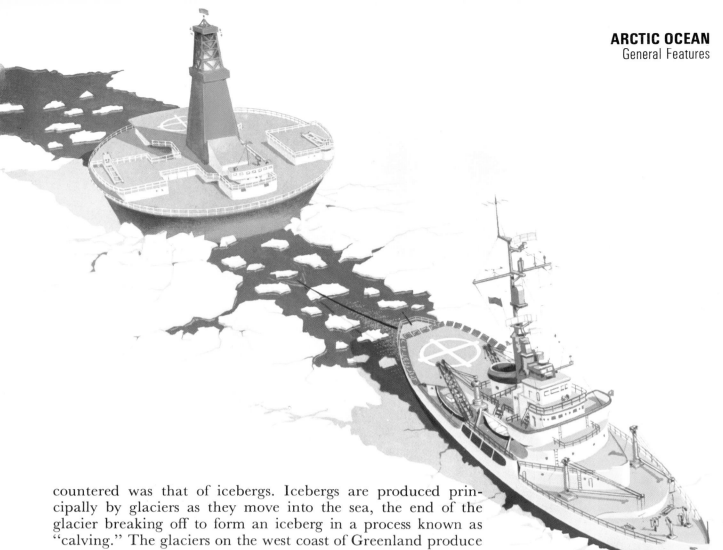

An icebreaker tug tows an oil rig, the hull design of which utilizes a concept adopted by Fridtjof Nansen in the 1890's for his Arctic exploration ship Fram. *The design incorporates a cone-shaped hull that, like* Fram, *would be squeezed upwards by pressure of ice forming around it. Eventually, so much of the hull would be out of the water that the weight of the unit would break the ice and it would settle back again.*

countered was that of icebergs. Icebergs are produced principally by glaciers as they move into the sea, the end of the glacier breaking off to form an iceberg in a process known as "calving." The glaciers on the west coast of Greenland produce more icebergs than any others in the Northern Hemisphere, since some of them move at rates of up to 90 feet a day. Greenland's Jacobshavn glacier, for example, calves over 1,000 icebergs a year, and these move out into the open sea to drift south towards the Atlantic Ocean. Some of them are huge, many reaching over 200 feet in height above the water, while some have been recorded at heights of 500 feet.

Working in often frightening conditions, scientists and engineers have been devising ways of coping with the continual threat posed to drilling operations by icebergs, although it is often impossible to predict their exact course. Sometimes one that is estimated to be passing the rig at a safe distance takes a sudden lurch because of the vagaries of deep currents, and bears down on a rig. This calls for swift and drastic action. In the case of the *Typhoon* program a tug and two supply boats stood by during drilling, and when a 500,000-ton iceberg came too close the tug's towing wire was run right around the berg so that it could be towed out of harm's way.

Pack ice in the Arctic Ocean has an average thickness of 10 to 14 feet, although it can be considerably more in many areas because of buckling and folding caused by the movement of floes in the slow currents. So the tendency is for objects in the path of the ice to be slowly crushed. One solution to this problem has been thought out by the giant US corporation General Dynamics and this returns to Nansen's ideas, as embodied in the *Fram*. The General Dynamics design calls for a cone-shaped hull to the drilling unit which, like *Fram*, would

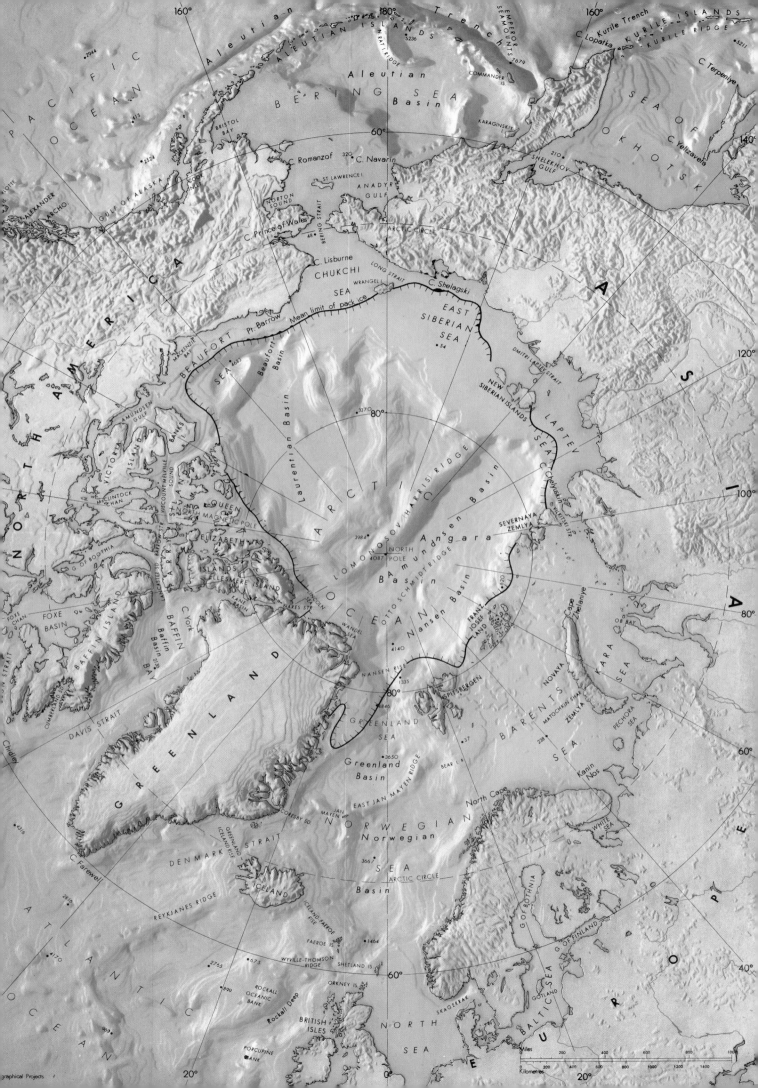

be squeezed upwards by ice pressure. Eventually, so much of the hull would be out of the water that its very weight (11,000 tons) would break the ice and it would settle back into the sea.

Once oil has been discovered in commercial quantities there is still the problem of getting it to the consumer. And this brings us to the other two Arctic voyages of discovery that in some ways were just as exciting, and of equal significance to the future of Arctic exploitation, as that of *Fram*. In 1969, a group of oil companies led by Humble Oil, investigating ways of bringing to market the massive discoveries on Alaska's North Slope, set out to discover whether it would be possible for supertankers to crunch their way through the Northwest Passage to Prudhoe Bay, Alaska, on a year-round basis. To test their theories they used the United States' biggest merchant vessel, the 115,000-ton, 1,000-foot-long tanker *Manhattan*.

Fitted with a massive, ice-breaking bow, and especially strengthened in critical areas, *Manhattan* took from September to November 1969 to make the return journey through the Northwest Passage. The 10,000-mile journey was tough: on occasions the ship had to feed all available power to the engines in order to break through the encircling ice. This meant switching off all lights, heating, and auxiliary machinery, backing the vessel up and then charging into the ice in order to punch a way through. The ship did not emerge unscathed: twice, holes were ripped in the hull, but not in critical areas. As a result of this and a later cruise, the feasibility of year-round tanker operations in the Northwest Passage has been proved. Although the oil from Alaska's North Slope will initially be transported by the ecologically controversial land pipeline, the discovery of offshore oil could see supertankers being used in the area.

As well as land pipelines and ice-breaking supertankers, there is another form of Arctic oil transport under active consideration that could have profound effects on the future of world sea transport as a whole. The argument is straightforward enough: instead of battling through freezing winds, fog, pack ice, and icebergs, why not take the cargo below the ice, leaving many of the problems of Arctic operations behind?

The idea stems from experience gained on the third major voyage of Arctic exploration to be described in this chapter. On August 3, 1958, the United States Navy nuclear submarine *Nautilus* passed under the 25-foot thickness of ice that marked the top of the world—the North Pole. It was the submarine's third attempt in a year and it made its way from the Pacific, through the Bering Strait into the Chukchi Sea, veered over towards Point Barrow in Alaska, and then sailed in a straight line under the ice towards the Pole. *Nautilus* was under the ice for 96 hours and emerged north-east of Greenland.

The voyage of *Nautilus* and subsequent subpolar trips by the nuclear submarines *Skate, Sargo,* and *Seadragon* have opened up vast new possibilities for subsea transport, not only of oil but of other bulk commodities. Merchant nuclear submarines making a transpolar journey from Japan to Europe could save 4,900 miles over the traditional sea route. Already, General Dynamics has proposed 17,000-ton supertanker-submarines to

The pioneering subpolar journey of USS Nautilus *could open up a whole new era of marine transport in which submarines would sail under the Arctic, cutting thousands of miles off the traditional sea route from, say, Japan to Europe (as shown on the map below). The submarines would use sonar that "looked" forward, up, and down to help them navigate under the ice.*

transport the oil from Alaska's North Slope to North Atlantic ports. Eighteen submarines, each 900 feet long, could carry oil from Alaska to Greenland, Iceland, or even Europe, along routes opened up by the pioneering US Navy submarines. The two-way journey time from North Slope to Greenland, including loading and unloading, would take about two weeks; even a trip from Alaska to Tromso in Norway and back would take only slightly longer—the distance (one way) is only 2,731 miles. The alternative journey would be through the Northwest Passage across the Atlantic—a distance of over 4,500 miles.

Thus, through the discovery of oil, man's knowledge of the hostile Arctic Ocean is increasing at a tremendous rate. Far more has been learned about the movement of sea ice in the Arctic Ocean, ice thickness, winds, waves, and currents in the past 15 years than over the preceding century. Before too long this knowledge will be put to use to make the Arctic Ocean as much utilized as the Atlantic.

6 The Southern Ocean

Huge icebergs, cold water rich in animal life—particularly whales—and swirling currents characterize the massive ocean of the Southern Hemisphere.

In the absence of land masses the ocean surface currents between the tropics would tend to be driven by the trade winds right around the world in roughly an east-west direction. Only the obtrusion of land masses breaks up this flow, combining with the density and wind distribution patterns to produce vast whirls and counter-currents. Between mid-latitudes and the poles the westerlies tend to drive the currents back to the east. In the Northern Hemisphere the west to east flow is modified by land masses. In the Southern Hemisphere, however, there is no land mass of significance in the vast area between 40° and 65° South. The westerlies thus drive the water before them in a globe-encircling current of immense power. Known as the Antarctic Circumpolar Current, it moves as much as 3,000 million cubic feet of water *every second* where there is nothing to obstruct its path.

But there is a bottleneck. The tip of South America stretches out towards the South Shetland Islands and the Antarctic Peninsula, leaving a gap of only 500 miles called Drake Passage. Here, the Circumpolar Current squeezes through the bottleneck and into the Scotia Sea, transporting a tremendous 5,300 million cubic feet of water every second—well over 600 times the volume of water carried by the Amazon. Winds, too, are fierce, with intense meteorological depressions moving across the Passage. Small wonder that sailors, from Sir Francis Drake to Sir Francis Chichester, have had cause to fear Cape Horn.

The Circumpolar Current does not flow in complete isolation from the rest of the global surface system. Undersea ridges and projecting land catch and twist it, sometimes breaking this vast oceanic river up into streams that loop away from the direction of the main stream. For example, where the current

Wires part and tangle on this ship caught in a storm off Cape Horn. The gap between the Horn and the Antarctic Peninsula (called Drake Passage) is only 500 miles wide. It represents a bottleneck to the Antarctic Circumpolar Current, which would otherwise flow unobstructed round the world between 40° and 65° South. It is an area, too, of intense meteorological depressions, and winds whip up the sea to frightening heights.

reaches South America (and from a glance at the map one could imagine Tierra del Fuego almost to be wincing away from the impact) a tributary is driven up the west coast to form part of the Peru Current.

In addition, cold Antarctic water is driven north by the west wind and its own relatively high density. Where it meets warm water drifting south from the Atlantic, Pacific, and Indian oceans, at roughly 50°s, the colder, denser Antarctic water sinks below the warmer, less dense southward-flowing water at what is called the Antarctic Convergence. This vast mid-ocean waterfall is invisible, but its effects become quickly apparent to anybody sailing south across it. In the Northern Hemisphere, the change from, for example, subtropical to temperate conditions is felt gradually. But the Antarctic Convergence acts like an invisible wall. Cross it and the air chills, the skies grey, and sea temperature drops markedly. The Convergence forms a biological boundary, too: many species of fish, birds, and plants found in abundance on one side of the boundary are rarely seen on the other. A further, more variable boundary occurs farther north (at around 40–45°s) where the now warmer subantarctic water meets subtropical water at the Subtropical Convergence.

Sail farther south from the Antarctic Convergence and you meet ice. Compared to the Antarctic the vast quantities of ice

The forbidding Antarctic landscape. Some 90 per cent of the world's snow and ice lies in Antarctica. Melt it, and the oceans of the world would rise by about 200 feet.

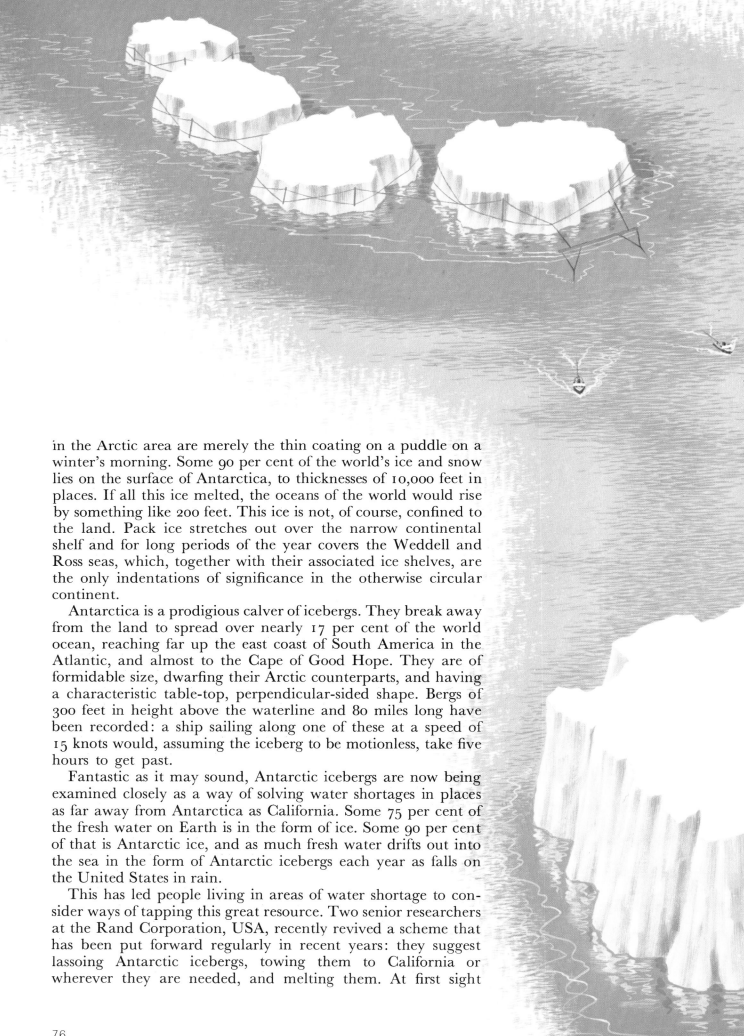

in the Arctic area are merely the thin coating on a puddle on a winter's morning. Some 90 per cent of the world's ice and snow lies on the surface of Antarctica, to thicknesses of 10,000 feet in places. If all this ice melted, the oceans of the world would rise by something like 200 feet. This ice is not, of course, confined to the land. Pack ice stretches out over the narrow continental shelf and for long periods of the year covers the Weddell and Ross seas, which, together with their associated ice shelves, are the only indentations of significance in the otherwise circular continent.

Antarctica is a prodigious calver of icebergs. They break away from the land to spread over nearly 17 per cent of the world ocean, reaching far up the east coast of South America in the Atlantic, and almost to the Cape of Good Hope. They are of formidable size, dwarfing their Arctic counterparts, and having a characteristic table-top, perpendicular-sided shape. Bergs of 300 feet in height above the waterline and 80 miles long have been recorded: a ship sailing along one of these at a speed of 15 knots would, assuming the iceberg to be motionless, take five hours to get past.

Fantastic as it may sound, Antarctic icebergs are now being examined closely as a way of solving water shortages in places as far away from Antarctica as California. Some 75 per cent of the fresh water on Earth is in the form of ice. Some 90 per cent of that is Antarctic ice, and as much fresh water drifts out into the sea in the form of Antarctic icebergs each year as falls on the United States in rain.

This has led people living in areas of water shortage to consider ways of tapping this great resource. Two senior researchers at the Rand Corporation, USA, recently revived a scheme that has been put forward regularly in recent years: they suggest lassoing Antarctic icebergs, towing them to California or wherever they are needed, and melting them. At first sight the scheme looks like a joke. But the two Rand research workers, John Hult and Neill Ostrander, are convinced of its feasibility. They say that a two-square-mile table-top iceberg will yield a million acre-feet (43,560 million cubic feet) of water. If several of these were linked together and towed like a train as much as 10 million acre-feet (435,600 million cubic feet) of water could be brought north at one time. The ice would have to be towed slowly—a speed of about one knot is proposed—and so the journey from Antarctica to southern California would take about 10 months. Each berg would need to be wrapped in double sheets of plastic film, separated by water, to make a quilt so that it would not be eroded by the seawater flowing past it. Hult and Ostrander say that it will not need protection from sun and air, and if shielded from seawater erosion each iceberg would lose only a tenth of its water during the ten-month voyage. On arrival, the icebergs would be moored, broken up, and the pieces fed into flexible underwater pipes, to arrive on shore in the form of fresh water or ice slurry. And the cost? Some 2½ times cheaper than southern California pays for some of its water supplies at present.

Ecology, "natural environment," "over-population," and other similar expressions have come into everyday use only during the past few years and are often used in an emotive rather than a scientific context. But they have been around for a

Researchers are looking at the feasibility of lassoing Antarctic icebergs (which are practically all fresh water) and towing them to California, where they would be broken up and the pieces pumped ashore to augment the fresh water supply—an idea that sounds fantastic, but is being treated seriously in America. The journey from the Antarctic to California would take about 10 months, and if the icebergs were shielded from the effects of seawater erosion only one tenth of the fresh water they represent would be lost during the voyage.

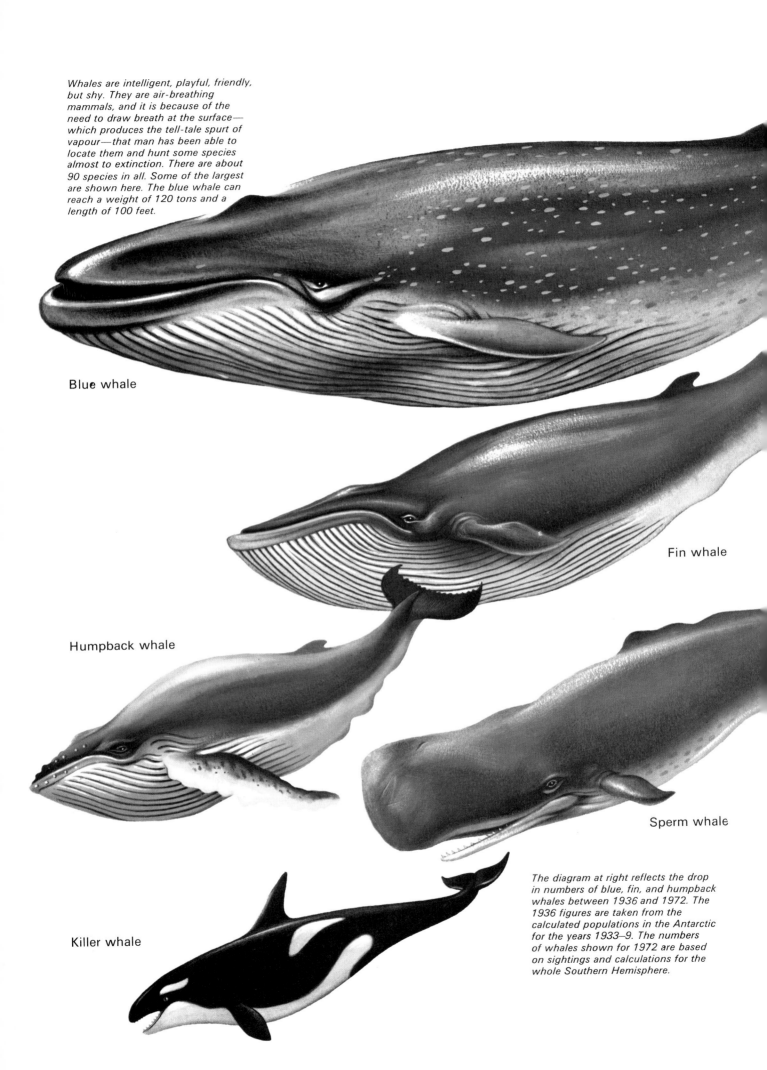

Whales are intelligent, playful, friendly, but shy. They are air-breathing mammals, and it is because of the need to draw breath at the surface—which produces the tell-tale spurt of vapour—that man has been able to locate them and hunt some species almost to extinction. There are about 90 species in all. Some of the largest are shown here. The blue whale can reach a weight of 120 tons and a length of 100 feet.

Blue whale

Fin whale

Humpback whale

Sperm whale

Killer whale

The diagram at right reflects the drop in numbers of blue, fin, and humpback whales between 1936 and 1972. The 1936 figures are taken from the calculated populations in the Antarctic for the years 1933–9. The numbers of whales shown for 1972 are based on sightings and calculations for the whole Southern Hemisphere.

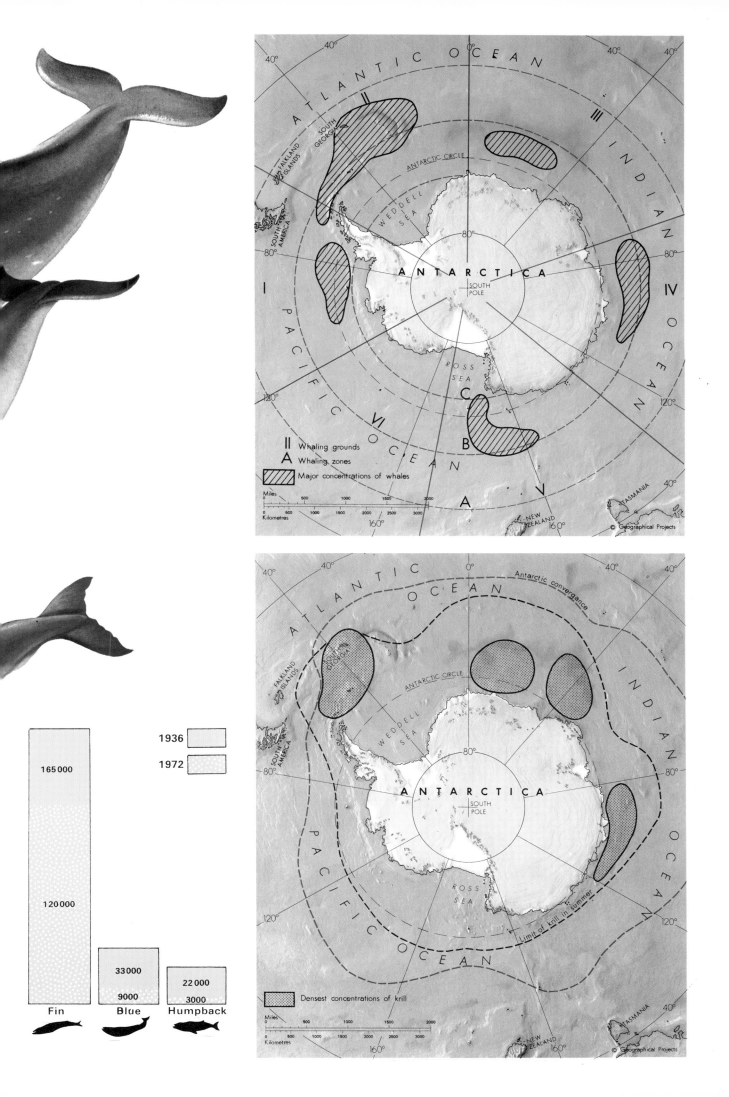

long time so far as Antarctic whales are concerned. The systematic depletion of stocks of whales to the verge of extinction—on the basis of "profit today and who cares about tomorrow"—provides an object lesson that is only today being appreciated by those who until now have taken the world's riches for granted.

Whales can be divided into two main groups: baleen (or whalebone) and toothed. The baleen whales include the largest mammals known to man. The blue whale, for example, reaches a length of 100 feet and weighs 120 tons. But for all their size these creatures are harmless. No giant teeth and jaws: instead, a feeding method which is more like a mighty vacuum cleaner. A sieve of fine plates hangs from the roof of the mouth and the animal swims through shoals of small shrimp-like creatures known as krill, sieving millions of them from the water.

Every year, around autumn, baleen whales migrate from the Southern Ocean to warmer, subtropical waters, where they breed. This process of migration is still not fully understood. It is known that the whales do not depart from the Antarctic *en masse*, but rather in a long procession, spread over a period of time. The exact location of breeding grounds, too, is not clearly

Hacking whale flesh on a factory ship. Many of the hundreds of products that used to be produced from whales are no longer required, or substitutes have been found for them. The United States has placed a total ban on the importation of whale products.

known. Experts will commit themselves no further than to say that breeding takes place in deep, subtropical water in latitudes as far north as 20°s. During the breeding season baleen whales are thought to eat little or nothing, living mainly off the thick layer of fat, or blubber, which they have accumulated under their skin while living on the Antarctic feeding grounds.

Sperm whales, the largest of the toothed whales, also migrate, but again, little is known of the exact pattern, except that only the males migrate, and in the opposite direction to the baleen species. Sperm whales live in warmer, subtropical waters and the males move down to the colder latitudes once a year. This pattern could well be related to feeding habits: sperm whales, like their smaller cousins the dolphins, are toothed animals and eat fairly large creatures such as squid and fish, which they may follow down to the Southern Ocean.

Whales are sociable, amiable creatures. The smaller whale species such as dolphins are now well known to be fond of play for play's sake, to go to the assistance of one of their number when it is injured or ill, to have remarkable intelligence, and a sophisticated system of communication that takes the form of whistling and clicking. There is no reason to suppose that their larger counterparts do not have all these characteristics. There is, therefore, something of a cruel irony about the way in which these harmless, shy creatures have suffered so much at the hands of man, their greatest single enemy. They have suffered because of one trick—their Achilles' heel—that nature has played on them. Whales are air-breathing mammals—creatures that used once to live on land, but that have returned to the sea. Under normal, peaceful circumstances a large baleen whale swims a few feet below the surface, breaking it every five to ten minutes to breathe for a period of about 20 seconds. As it comes towards the surface the whale starts to exhale through its nostrils, which are on top of its head, sending out a powerful spurt, or "blow," of vapour high into the air. This simple act of breathing spells death if there is a whale-catching boat anywhere in the vicinity, for it is the only way by which a whale can be spotted at a distance.

That revealing spout has supported an industry that was a high-seas activity in the 1400's but which began in the Southern Ocean at the beginning of this century with the establishment of a land station on the island of South Georgia. From this station catcher boats operated within a range of 100 miles, using the harpoon cannon, which had been invented by a Norwegian, Sven Foyn, in 1860. The 1925–26 season, however, saw the introduction of the factory ship operating with its retinue of catching boats, and capable of hauling the entire whale carcass up a slipway and onto the deck. This ushered in the era of "pelagic" (to distinguish from land-based) whaling and soon made the Southern Ocean the most important whaling area in the world.

The impact of these ships was tremendous. In the 1930-31 season 41 vessels operated with 200 catcher boats in attendance on them. They established a record that will probably never be broken: a total of 37,465 whales, of which no fewer than

28,325 were blue whales. For whale production to have continued at anything like that of the 1930-31 season would have soon meant extinction of the main species. As it was, the enormous catches that season knocked the bottom out of the market for whale products and the industry made voluntary, self-imposed production agreements for the following seasons.

But this was not enough. Scientists feared for the extinction of some species, and gradually still further restrictions were imposed, including a total ban on the taking of gray and right whales, which began in 1946. In 1946, too, the International Whaling Commission, which is headquartered in London, was established. Made up of representatives from each government that is a party to the International Whaling Convention, the Commission regulates whaling throughout the world, deciding, among other things, an annual total catch limit for Antarctic pelagic whaling.

The Commission set this limit in terms of blue whale units (BWU), each unit equalling one blue whale—or two fin whales, $2\frac{1}{2}$ humpbacks, or six sei whales. It seemed a good idea at the time, but it nearly sounded the death knell for the large baleen whales. The Commission itself admitted in 1972, the year when the system of blue whale units was dropped, that "adoption of the blue whale unit was the principal cause of the decline of the Antarctic whale resources."

What went wrong was that the whalers, having been given their BWU quota, immediately set out after blue whales, because a single blue was at least twice as valuable as any other. The stocks of blue whales were quickly decimated. Then the whalers turned to fin whales—the next most valuable species—then to humpbacks, then to sei. The plight of some species became desperate. In 1963 the total population of blue whales in the Antarctic was down to between 1,000 and 3,000. Humpback whales, too, were in danger of being hunted out of existence and since 1966 both they and blue whales have been subject to strict conservation methods in order that they may survive.

The 1972 United Nations Conference on the Human Environment in Stockholm came to the defence of the remaining whales by demanding a complete moratorium on whaling for 10 years. This resolution, tossed into the collective lap of the International Whaling Commission when it met shortly after the Stockholm conference, resulted in one of the most productive sessions in the Commission's history. The moratorium was rejected, but total protection was extended to blue and humpback whales; the blue whale unit was abandoned in favour of a catch quota by species. Perhaps most important of all, the Commission received a report that an International Observer Scheme had been agreed by the Antarctic whaling nations of Japan, USSR, and Norway, which assures an exchange of observers between their factory ships operating in the Antarctic.

The imposition of catching restrictions and the decline in stocks has, of course, led to a corresponding decline in whaling effort in the Antarctic, so that in the 1971-72 season just seven expeditions operated there, four from Japan and three from the USSR.

Some familiar animals of the Southern Ocean: southern elephant seals (the male has the characteristic trunk-like nose), Adélie penguins (at left of group), an emperor penguin, and a silver-grey fulmar.

The largest mammals on earth have thus been saved—but only just. Stocks of the large Antarctic whales are now beginning to build up again (the number of blue whales is today put at around 10,000) and so one day it should be possible to start culling them again.

But will they ever need to be exploited on the same scale as in the past? Whalebone—the fine plates that hang from the roof of the mouth of the baleen whales—is no longer required for use in ladies' corsets, or to provide springs for carriages. Sperm whale oil is no longer needed for oil lamp fuel; and it is rapidly being replaced by synthetic substances in the production of soap and cosmetics. The oil from baleen whales is edible and is used to manufacture margarine, cooking fats, and soap, but vegetable substitutes are finding favour. Even the use of whalemeat for human and animal food could perhaps also be abandoned by going one step back in the krill-whale-man food chain and perfecting methods of catching and processing the baleen whales' chief source of food. Krill occur in vast quantities in the Antarctic and already Soviet factory ships are experimenting with methods of catching and processing them to make a type of fish paste containing 24 per cent protein.

With the emphasis today on a better understanding of and respect for our environment, we must hope that man will in future learn to live without the ruthless exploitation of these huge creatures. The United States has set a lead here and shown that it is possible to live in harmony with whales by placing a total ban on the importation of whale products.

But even if the killing must continue, surely a more humane method can be devised? Imagine the outcry if cows were killed by having a flesh-ripping explosive harpoon slammed into them, so they died in agonized shock. Whalers argue that there is no quicker method of killing the animal—providing the first shot is well placed. But the whole barbaric process is thought by many to be one against which the full armoury of modern technology should be enlisted. If we have to wage war on whales, at least let us make sure that the battle is an honourable one.

7 Far Eastern Seas

A bewildering variety of form and activity, from the icy waters of the Bering Sea to the warm seas around the East Indian Archipelago.

From the five great oceans we turn to the seas. All the world's seas are interconnected parts of the global oceanic system. They fall into two main categories: adjacent, or marginal, seas are those connected individually to the larger water body; mediterranean seas are groups of seas forming an individual unit collectively separated from the main water body.

Both types occur in the bewilderingly complex puzzle of seas that stretch down the western side of the Pacific and around to the Indian Ocean. The Philippine Sea is an example of an adjacent sea; farther south, the seas of the Indonesian archipelago form a mediterranean complex.

The Far Eastern Seas are bordered to the west by the land mass of the USSR, China, Indo-China, and Malaysia, and to the south by the East Indian Archipelago. (Some authorities also include the Arafura and Timor seas, which lie between the Archipelago and Australia.) To the east the seas are separated from the Pacific Ocean by a system of trenches and mid-ocean ridges. The ridges are topped by islands such as the Aleutians, Kuril, Japan, the Ryukyu Islands, Taiwan, and the Philippines. Off Japan the ridges part and meet again south-east of the Philippines, having formed a roughly oval shape that encloses the vast Philippine Sea.

Because they stretch from near the Arctic Circle to south of the equator the Far Eastern Seas offer a contrasting variety of features. Many of them are by no means the shallow, sandy basins that might be expected by those familiar with, say, the North Sea. The Sea of Japan, for example, has an average depth of nearly 4,500 feet and is over 13,000 feet at its deepest point. It is furrowed by deep trenches, alternating with large banks that rise to within a few hundred feet of the surface, giving the

A Chinese junk under sail. Although the Far Eastern Seas are usually associated with the complex pattern of water and islands in the area from Japan to Indonesia, in fact they stretch from the Arctic to south of the Equator.

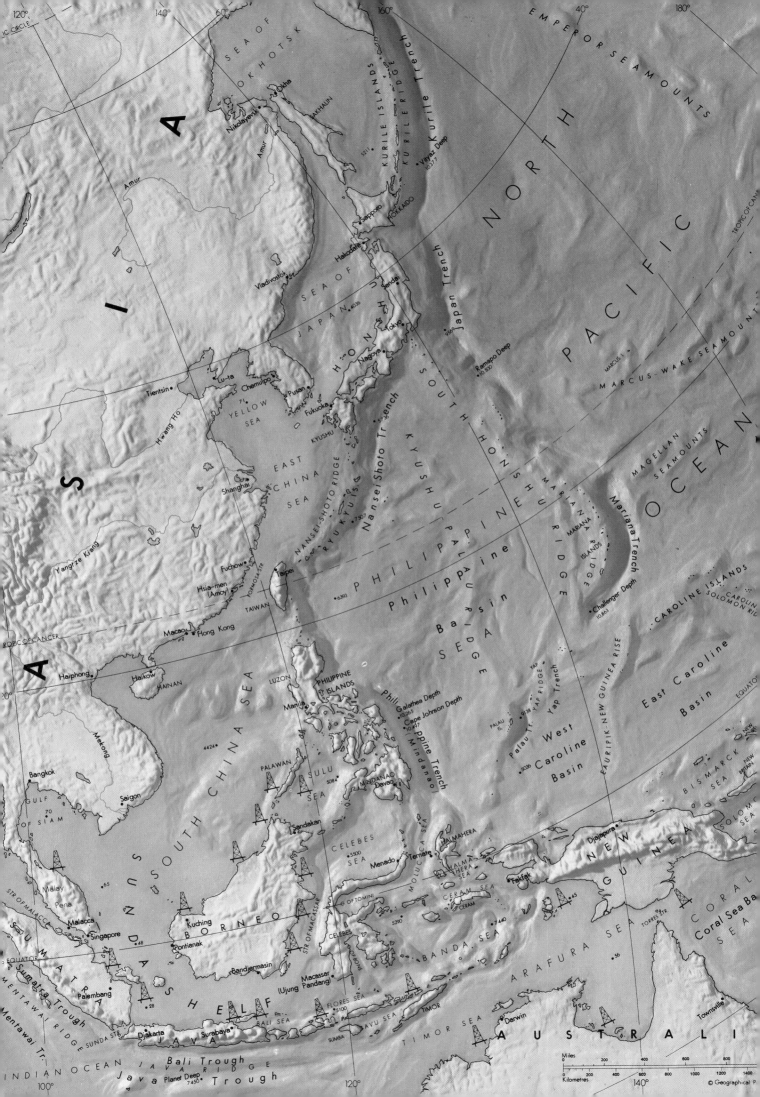

Fishing boats at work in the near-Arctic conditions of the Bering Straits—the northernmost of the Far Eastern Seas. The waters of the Bering Sea are rich in demersal fish such as cod, halibut, and flatfish, as well as pelagic species such as salmon and herring, and a cosmopolitan fleet searches for them, coming from the United States, Canada, Japan, and the USSR.

The bewildering pattern of the Far Eastern Seas. As in nearly every other area of the world continental shelf, the search for oil and gas is intensifying, and the main areas of exploration are shown on this map.

whole sea an awesome underwater landscape. But farther south the Java Sea, part of the South China Sea, and the Gulf of Thailand cover one of the largest continental shelves in the world. This is the Sunda Shelf, and it occupies some 700,000 square miles, with most of this area lying under less than 150 feet of water. In fact a million years ago the Shelf was exposed, so that Malaya, Sumatra, Java, and Borneo were once part of a single land area.

The Far Eastern Seas offer every extreme of climate, too. To the north, in the lonely Bering Sea where there is little shipping, winter temperatures seldom rise above freezing point, and Russian and Japanese fishing vessels have often to nose through pack ice in search of the herring, salmon, cod, halibut, redfish, flatfish, and whales that are plentiful in the area. To the south, in the bustling Philippine archipelago, the crews of large and small vessels learn to judge the moods of the monsoon as they ply between the islands or steam to one of the major Far Eastern ports, such as Hong Kong or Singapore. The local fishing boats search for colourful tropical fish—tuna, sardines, scads, and pompanos; slipmouth, croaker, and shrimp; barracudas, snappers, and groupers.

Between these extremes of distance, climate, and topography one pervading influence provides a common link: that of Japan. If any nation needs to make the fullest use of the oceans it is Japan. Her islands have a land area half as big again as that of the United Kingdom, but only 16 per cent of the total is suitable for agriculture, industrial development, and housing.

And so Japan has turned to the sea. We have seen already how over half of Japan's animal protein requirement comes from fish products, and her fishing vessels, from small, two-man trawlers to factory ships, work throughout the Far Eastern Seas catching anything from sardines to whales. Other basic commodities come from the continental shelf of the Japan Sea and the Pacific: huge artificial islands built off the coast mine 12 million tons of coal a year; ore-rich sands are dredged for the iron they yield. Japan is the largest importer of oil in the world, relying on foreign sources for 99 per cent of her supplies. In an effort to reduce this reliance she has initiated a vigorous exploration policy and along with United States companies is drilling on the Sunda Shelf off Sumatra, Java, Kalimantan (Borneo), and Celebes.

Japan's dominance of the world shipbuilding industry is well known. And from those small islands her own ships thread their

Japan's land shortage and lack of indigenous natural resources have meant that she has turned increasingly to the sea for food, fuel, minerals, and even living space. Already she dominates world ship-building. Here, a ship is launched with the usual colourful display of streamers and balloons at the bow.

89

way through the Far Eastern Seas to take exports to world markets, making the Malacca Straits (between Malaya and Sumatra) as busy as the English Channel.

But perhaps the most exciting of all Japan's marine activities are her efforts to extend living and working space by moving into the sea. The nation's continental shelf has an area of 108,000 square miles—about three-quarters of the land area. Of this, nearly 70 per cent lies under less than 165 feet of water and so represents a challenge to Japanese technology to colonize it.

The general plan is to take land-consuming and polluting industries and site them on the continental shelf. So, if power stations take up valuable land space, disfigure the landscape, and pollute coastal waters, move them out to sea. There are plans for 20 conventional and nuclear power stations to be built offshore; some on fixed platforms in shallow water, some floating in deeper water, and others resting on the seabed, connected by cable or tunnels to the shore. There is an acute water shortage so desalination plants are to be built alongside the floating nuclear power plants.

Too little room for bulk storage of oil, grain, fish meal, and other materials? Keep them in underwater storage tanks, perhaps rather like those in the Persian Gulf, where low, even temperatures underwater will often be of help in keeping commodities fresh. No room for the expansion of heavy industry? Man-made islands up to three square miles in area provide space and cut land pollution. No room for another airport? Then build a 1,350-acre offshore island and connect it to the land by tunnel. No land on which to raise cows and sheep? Then farm fish instead. Japan already practises fish farming on a large scale, and if just five per cent of the total continental shelf area could be made over to farming, an extra six million tons of fish a year would be available.

These are just a few of the plans, some of which are already being put into effect. And the planners have not lost sight of the fact that the sea is a favourite place for relaxation and recreation. An underwater park, the first of many, has been built which features an underwater observation tower, an underwater restaurant, and a moving underwater walkway on the seabed inside a watertight tunnel. There are even plans for complete underwater cities, although whether Japan's overcrowding problem has reached such a stage that anyone can be persuaded to live in them is a matter for debate.

Japan is of necessity leading the world in making intelligent use of the extra territory available on her continental shelf. What she does today other nations will have to do tomorrow as the tiny part of this planet on which we live at present becomes just too small for us all.

This is an underwater observation tower off the coast of Japan. Access is by walkway from the shore, and an elevator serves the underwater section, where powerful lights allow the visitor to observe marine life.

8 The Caribbean and Gulf of Mexico

The Caribbean has an immense food potential for the surrounding island chain; the Gulf was the cradle of the offshore oil and gas industry.

Like the Far Eastern Seas, the Caribbean Sea is bounded to the west by a land mass and to the east by an arc of islands, subsea ridges, and trenches. The Caribbean's southern boundary is formed by the northern coast of the South American countries of Venezuela and Colombia, and by the Isthmus of Panama, with the Panama Canal providing a threadlike link between the Caribbean and the Pacific Ocean. Honduras and Yucatan form the eastern boundary and the narrow gap between Yucatan and Cuba, the 137-mile-wide Yucatan Strait, leads into the Gulf of Mexico. Together the Gulf of Mexico and the Caribbean Sea make up the American mediterranean.

The floor of the Caribbean Basin is a complex structure of ridges and trenches that cut across the sea's area of over one million square miles. The sea's greatest depth is 23,000 feet, in the Cayman Trench between Cuba and Jamaica. Other basins of great depth include the Grenada Basin (over 9,000 feet), the Venezuela (over 16,000 feet), Colombia (over 13,000 feet), and the Yucatan (over 16,000 feet).

Some of the deep trenches leading from these basins pass between the islands of the Antilles Chain into the Atlantic Ocean, those of sufficient depth and width—notably the Windward, Anegada, Dominica, Jamaica, and South Aves passages—playing a large part in funnelling water between the Atlantic and Caribbean. However, most of the passages between the islands and the open Atlantic are of insufficient depth to permit intermixing of anything but the surface layers of water. Consequently the movement of water at depths greater than about 3,500 feet is positively sluggish compared with that of the oceans, and in some areas this has a profound effect on the composition of the water. Generally speaking, the chemical composition of seawater

Fire ravaged this self-elevating oil-drilling platform in the Gulf of Mexico off Louisiana in 1971, and took weeks to extinguish. Fires and hurricanes are two of the main hazards of offshore drilling in the gulf, where over 13,000 exploratory wells have been drilled.

is practically constant throughout the world, such is the efficiency of the oceanic circulation system. But the concentrations of dissolved oxygen, nutrients, and of trace metals necessary for biological activity are known to vary and where the circulation is impaired, as in the Caribbean Sea, the nutrient balance can change considerably. The water in the Cariaco Trench off Venezuela, for example, is anoxic, or stagnant, because the oceanic circulatory system is not replenishing it. There is hardly any dissolved oxygen and a consequent dearth of marine life. There are also big differences in trace metals: in open seawater above the sill of the trench, dissolved copper occurs in concentrations of 0.003 milligrams per litre—the normal level for open seawater; in the trench itself, however, the concentration is so low that usual analytical techniques cannot detect it. Strangely, the iron concentration in the trench is a thousand times greater than in the open ocean.

In contrast to the Caribbean Sea the Gulf of Mexico has a relatively simple structure, following almost an idealized pattern of broad continental shelf falling gently to a continental slope that sinks sharply to a small abyssal plain, which at its deepest is just over 12,000 feet deep. The continental shelf is at its widest off Florida, then gradually narrows along the coast of Mexico until it reaches the Yucatan Peninsula, where it again broadens.

There are no trenches or ridges of any significance in the Gulf of Mexico, but in the centre of the abyssal plain a series of hills, known as the Sigsbee Knolls, thrusts up to heights of 1,200 feet above the surrounding flat seabed. There was some debate about their composition until the drilling ship *Glomar Challenger* began her famous career there in 1968 by probing through the ocean depths to take samples from one of the hills. From analysis of the cores it was found that the hills were salt domes, topped by a caprock that, on one of the sites drilled, contained oil. This was the first time that oil had been found at such depths, and the discovery offers exciting possibilities for the future. But

The gulf is also a marine playground for sports fishermen who go in search of the grotesque hammerhead shark and the graceful tarpon, shown here.

Gulf of Mexico shrimp boats rest at anchor. The gulf supports the biggest shrimp industry in the world, and catches reach an incredible 500 million pounds a year.

at the time, the discovery was something of an embarrassment to the *Glomar Challenger* scientists who, worried about pollution and possible international legal entanglements (it has not yet been decided who owns resources on and beneath the deep-ocean floor), hurriedly plugged the hole they had made with cement to make it safe, and departed. But one can be sure that American oilmen working on the continental shelf of the Gulf of Mexico have carefully marked the spot where the discovery was made for future reference!

Oil and fisheries are the two marine activities that dominate the Caribbean Sea and Gulf of Mexico. In the Caribbean great efforts have been made in recent years to take more advantage of the rich seafood resources, particularly shrimps, that the area offers. The native population, its numbers swelled increasingly by tourists from America and Europe, are great fisheaters. Until now this appetite has had to be met largely by imports, but several ambitious projects aim to bring about not only a self-sufficiency in fisheries products, but to make the islands net exporters of fish.

One country has dramatically shown the rest of the Caribbean islands how this can be achieved: Cuba. In 1958 the Cuban fish catch was 21,000 tons. Today it is up around the 110,000 tons a year mark, and heading still higher as the Cuban government boosts fisheries investment as one of the ways to move the country away from its economic over-dependence on the sugar cane industry. The tremendous development of Cuban fisheries has been achieved with massive Soviet support, and the industry is now organized on centralized, Soviet-style lines.

Headed by the Instituto Nacional de la Pesca, the Cuban fleet is divided into five sections. La Flota Camaronera del Caribe (Caribbean Shrimp Fleet) consists of 400 trawlers that operate in the Gulf of Mexico and over the Caribbean. Most of the vessels have refrigeration and process on board the shrimp they catch. La Flota Camaronera del Sur (Southern Shrimp Fleet) has 113

CARIBBEAN AND GULF OF MEXICO
General Features/Fisheries/Oil and Gas

Anchovy · Hake · Grouper · Menhaden

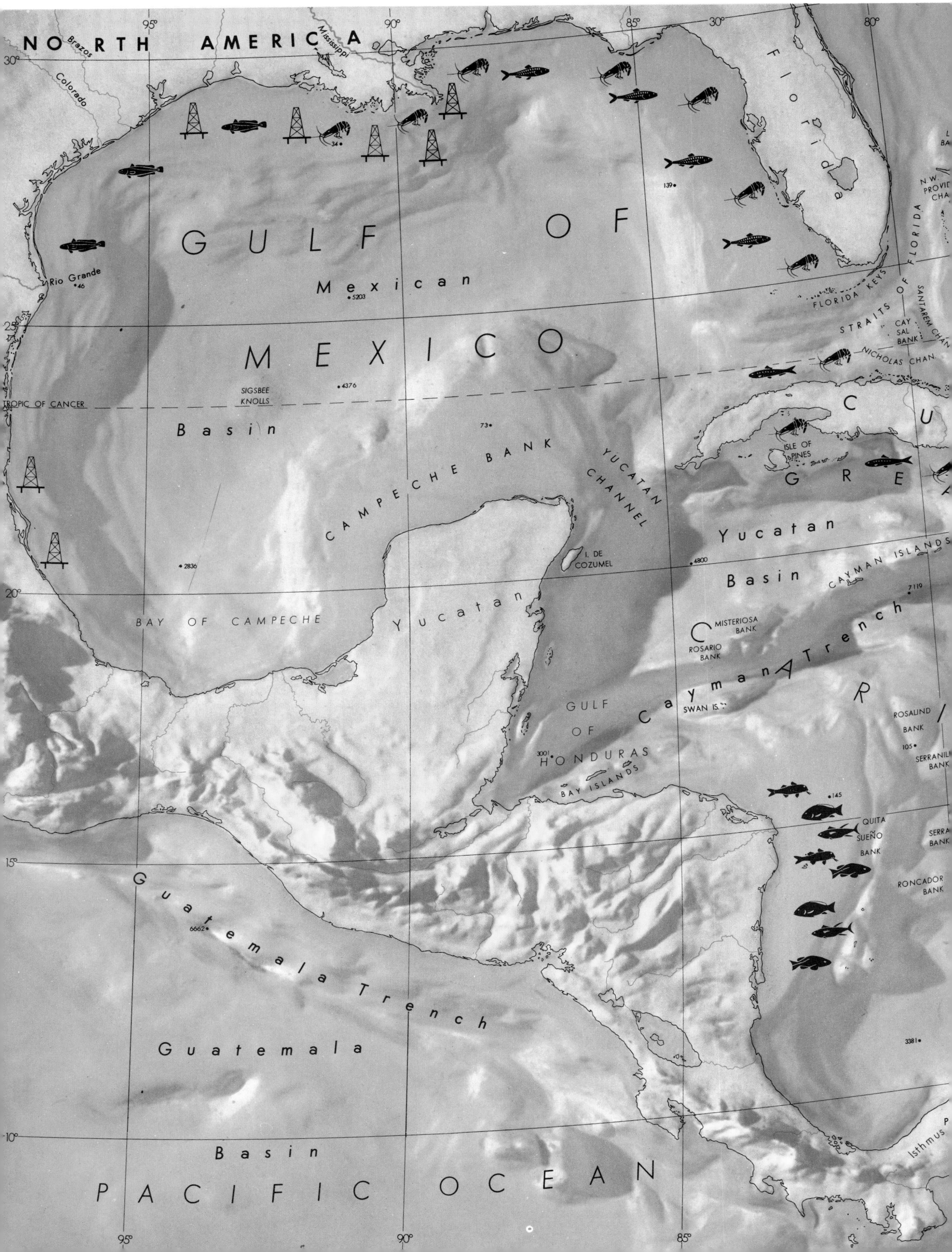

NORTH AMERICA

Brazos · Colorado · Mississippi · Florida

GULF OF MEXICO

Rio Grande

Mexican Basin

SIGSBEE KNOLLS

TROPIC OF CANCER

CAMPECHE BANK

YUCATAN CHANNEL

FLORIDA KEYS · STRAITS OF FLORIDA · CAY SAL BANK · SANTAREM CHAN. · NICHOLAS CHAN.

ISLE OF PINES

CUBA · GREATER

Yucatan Basin

I. DE COZUMEL

CAYMAN ISLANDS

MISTERIOSA BANK
ROSARIO BANK

BAY OF CAMPECHE

Yucatan

Cayman Trench

GULF OF HONDURAS

SWAN IS.

BAY ISLANDS

ROSALIND BANK
SERRANILLA BANK
QUITA SUEÑO BANK
SERRANA BANK
RONCADOR BANK

Guatemala Trench

Guatemala Basin

PACIFIC OCEAN

vessels of which, despite the name of the fleet, only 37 catch shrimp while the others are engaged in catching fish. La Flota del Golfo (Gulf Fleet) consists of 88 mainly wooden vessels that catch over 8,000 tons of fish a year. A measure of the increased efficiency of the "new" Cuban fishing methods is that four years ago this same fleet consisted of 140 vessels catching only 5,000 tons a year. La Flota Cubana de Pesca (Cuban Fishing Fleet) shows most clearly the Soviet influence. Operating from a $40 million Russian-built base it is a sophisticated fleet of 12 factory ships, six Soviet-built side trawlers, 24 tuna longliners, and associated refrigerated transport and supply mother ships, one of them a converted passenger liner able to carry 2,000 tons of fish. The fleet roams the world, sometimes making eight-month voyages to South Africa, and has a trans-shipment base in Las Palmas in the Canary Islands. The fifth group, La Flota de la Plataforma, has this name because its fleet of 107 vessels fishes the "platform," or continental shelf, round the Cuban islands for rock lobster, spiny lobster (crayfish), and fish, including sardines, as well as collecting sponges.

The huge fishing effort, which also includes establishments for breeding fish and crustacea, enables the Cuban fisheries export agency, Caribex, to offer the world a range of often exotic frozen

Lake Maracaibo in Venezuela was the cradle of offshore drilling when, in the 1920's, rigs were mounted on wooden-piled platforms connected to the shore by boardwalks. Today, drill rig derricks thrust up from the lake as far as the eye can see.

and canned seafood, including rock lobster tails, headless shrimp, squid, tuna, red snapper, grouper, sword-fish fillets, shark fins, and herring.

The other Caribbean countries, especially those in the Lesser Antilles Chain, are determined to have some of this fisheries wealth for themselves. Fifteen countries have participated in a research project assisted by the United Nations Development Fund and the UN Food and Agriculture Organization to develop deep-water fishing for sea trout, grouper, and snapper. Three FAO research vessels have explored over $1\frac{1}{2}$ million square miles of the Caribbean, chiefly around Jamaica, Trinidad, and Barbados to identify fish stocks; hundreds of master fishermen have been trained in modern methods; and eventually, with help from the United Nations and the developed nations, a fleet of 400 entirely Caribbean-owned trawlers should be operated from Grenada, St. Lucia, St. Vincent, Antigua, Barbados, Tobago, and other Caribbean countries, backed by modern fish processing and distribution centres.

The emphasis here is on "Caribbean-owned," for many of the islands already support large modern fleets and processing facilities owned and operated by United States seafood companies. These American bases, scattered all over the Caribbean, are geared to satisfying the voracious American appetite for shrimp. Landings from the Gulf by United States vessels alone reached 230.5 million pounds in 1970. Add to this 48.8 million pounds caught in the Gulf and Caribbean by overseas-based United States vessels, and an additional 250 million pounds of domestic production by nations such as Mexico, Cuba, Honduras, Panama, El Salvador, Venezuela, Trinidad, and others, and the total yield from the American Mediterranean reaches 530 million pounds a year—the biggest shrimp fishery in the world.

The shrimp industry supports enormous ancillary activities of processing, boat-building, and equipment supply; one Florida yard, for example, has produced nearly 2,000 trim 75-foot shrimp trawlers over the past 25 years. All three major shrimp species, white, pink, and brown, are caught in the American Mediterranean in water depths of up to 120 feet. And although the industry has its fluctuations, the overall trend in catches is still upwards. The Gulf of Mexico and the Caribbean will continue to hold their position as the world's premier shrimp grounds for many years.

The vast continental shelf of the Gulf of Mexico on which the shrimp industry thrives is also the most productive offshore oil and gas area in the world. In fact the Gulf, and Lake Maracaibo in Venezuela, at the southern end of the Caribbean, were the cradle of the offshore oil industry. In the 1920's it became plain that onshore oil and gas discoveries in Venezuela and the Louisiana swamplands extended out into Lake Maracaibo and the Gulf, and so the oilmen set about getting their feet wet. They moved cautiously at first. The first offshore drilling was from rigs mounted on wooden-piled platforms, so close to the shore that they had direct access by means of boardwalks. In the 1930's true offshore drilling began with the development of pontoon barges that were towed to a previously prepared seabed site and then sunk into position so that the flooded pontoons sat on the seabed with the drilling platform above the water.

Improved designs of fixed platforms followed until the end of the 1940's, when mobile rigs were introduced that could be towed from site to site without the expensive process of dismantling that the fixed platform necessitated. These took the form of the jack-up platform, which consists of a barge carrying the drilling equipment and legs that rest on the seabed. On the way to the drilling site the barge floats with the legs kept in a raised position. At the site they are jacked down through the hull of the barge until they rest on, or penetrate, the seabed. Jacking continues until the barge hull is raised clear of the water, out of reach of the waves. Then came the drillship, which lays anything up to eight anchors to restrict its movements while drilling. Finally, semi-submersible drilling rigs were introduced. These are massive structures with pontoons at the base of buoyant legs so that the rig floats in deep water, the submerged pontoons and partly submerged legs being far enough below the surface to be free of the worst effects of waves. The structure is thus afforded a far greater degree of stability than a drilling ship. The latest breed of semi-submersibles has propulsion engines, enabling it to travel from site to site under its own power.

Rigs of all these types operate in the Gulf of Mexico today and have drilled over 13,000 wells. In mid-1972 the record was: 44 jack-ups, 4 semi-submersibles, 1 drillship, and 90 fixed platforms. Nearly all this offshore drilling activity in the Gulf of Mexico takes place off Louisiana, with a little off Texas. Still to be explored are the vast areas of continental shelf off Florida, Alabama, and Mississippi.

Drilling for oil and gas on the continental shelf of the Gulf is not so fraught with difficulties as it is in other parts of the world, notably the North Sea. The shelf slopes gently away from the land, so that there are about 127,000 square miles of US continental shelf in less than 600 feet of water. Most of the operations take place quite close to shore, with the most distant fixed platform only 40 miles from land. And the weather, with one violent exception, is generally good.

That one exception is the hurricane. "Hurricane" is the name given to any wind of Force 12—over 75 miles per hour. More specifically it refers to the tropical cyclones that build up out in the Atlantic and in the southern Gulf of Mexico and Caribbean from July to October, and which can wreak havoc in the islands and along the coast of the Gulf. When a hurricane is forecast, mobile rigs, wherever possible, move to port. On fixed platforms the crew shut down the oil or gas wells and depart, leaving the hurricane to take its toll. The principal damage to platforms comes not from winds but from the gigantic waves whipped up by the storm: with the winds raging up to 150 miles an hour, waves can reach heights of 60 feet, smashing into the platform until, in the most severe circumstances, it collapses. The cost of the damage is high: in 1965 Hurricane Betsy was responsible for a damage bill to the offshore operators of over $175 million.

This is a service capsule in which engineers, breathing ordinary air, are lowered to a seabed chamber that contains, also at atmospheric pressure, all the equipment for completing a subsea well. Developed by Lockheed Petroleum Services, this equipment will one day permit oil to be produced from water depths as great as 3,000 feet.

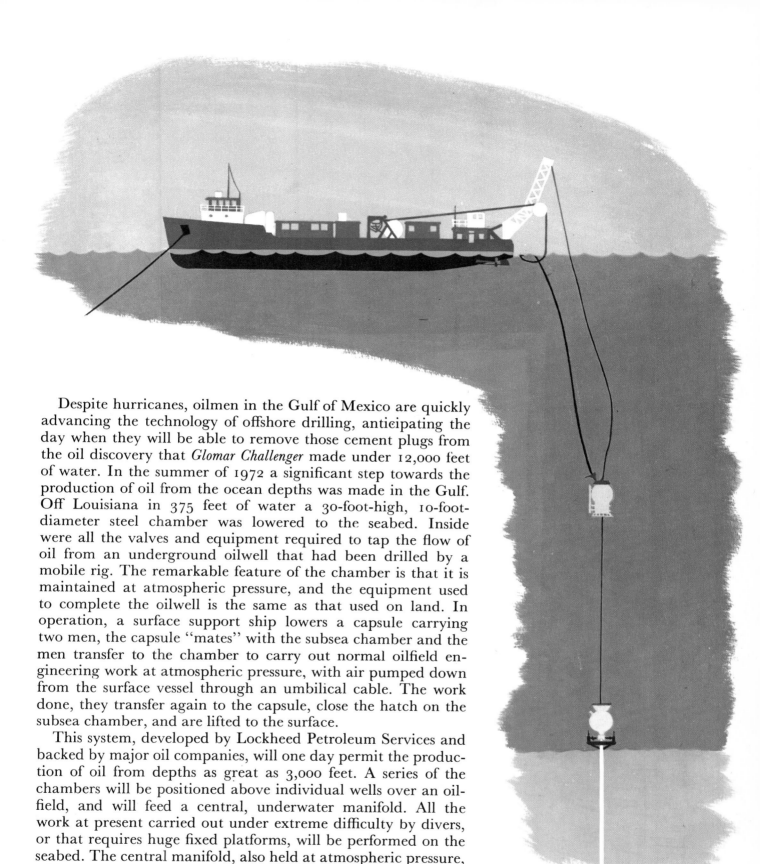

Despite hurricanes, oilmen in the Gulf of Mexico are quickly advancing the technology of offshore drilling, anticipating the day when they will be able to remove those cement plugs from the oil discovery that *Glomar Challenger* made under 12,000 feet of water. In the summer of 1972 a significant step towards the production of oil from the ocean depths was made in the Gulf. Off Louisiana in 375 feet of water a 30-foot-high, 10-foot-diameter steel chamber was lowered to the seabed. Inside were all the valves and equipment required to tap the flow of oil from an underground oilwell that had been drilled by a mobile rig. The remarkable feature of the chamber is that it is maintained at atmospheric pressure, and the equipment used to complete the oilwell is the same as that used on land. In operation, a surface support ship lowers a capsule carrying two men, the capsule "mates" with the subsea chamber and the men transfer to the chamber to carry out normal oilfield engineering work at atmospheric pressure, with air pumped down from the surface vessel through an umbilical cable. The work done, they transfer again to the capsule, close the hatch on the subsea chamber, and are lifted to the surface.

This system, developed by Lockheed Petroleum Services and backed by major oil companies, will one day permit the production of oil from depths as great as 3,000 feet. A series of the chambers will be positioned above individual wells over an oilfield, and will feed a central, underwater manifold. All the work at present carried out under extreme difficulty by divers, or that requires huge fixed platforms, will be performed on the seabed. The central manifold, also held at atmospheric pressure, will collect and control the output of oil from the various subsea chambers and will be able to pump it distances of several miles.

The tests now in progress off Louisiana with this new system are being eagerly watched by offshore oilmen throughout the world. If successful they could open up vast new reservoirs of energy, and, once again, the Gulf of Mexico will have been the cradle of an exciting step forward in man's underwater capabilities.

A boat lowers the service capsule to a seabed chamber on the floor of the Gulf of Mexico in 375 feet of water, so that engineers can check the equipment. The service capsule "mates" automatically with the seabed chamber. Eventually, a series of these chambers will feed a central manifold, which in turn will feed a subsea production complex.

9 The Mediterranean Sea

Rich both in history and for its exciting potential, this almost landlocked sea is threatened with choking death from pollutants.

During World War II the Allied navies were faced with a critical problem. German and Italian submarines were somehow getting into and out of the Mediterranean Sea and creating havoc among shipping. The Allies knew that the submarines could not be putting out from the ports in the Mediterranean or Black Sea: their only possible route could be through the Strait of Gibraltar. Yet all the might of the Allied navies could not catch the marauders. Careful listening for the tell-tale throb of the submarines' engines failed to reveal them; even the use of the newly developed Asdic, which pinged echoes off the steel hull of an underwater craft, was of no use. The submarines came and went with impunity.

Only after the war was the secret discovered. The submarine commanders were making use of a natural phenomenon that has a significant effect on the whole oceanographic make-up of the Mediterranean.

The average depth of the Mediterranean is nearly 5,000 feet, and there are extensive areas with depths of over 10,000 feet. But at the Strait of Gibraltar—the one outlet that prevents the Mediterranean Sea being entirely landlocked—an undersea shelf stretches across the Strait to within 1,000 feet of the sea surface. Because the 2,300-mile-long Mediterranean is surrounded by warm, dry countries, the amount of rainfall over the sea, even when combined with the runoff from rivers, is not sufficient to replenish the amount of water lost from the sea by evaporation. In fact, if the Mediterranean had to rely only on rainfall and rivers, its level would drop by 55 inches in a year. The "make-up" water that prevents this drop comes from the Atlantic Ocean, flowing in at the Strait of Gibraltar. Once into the Mediterranean the water flows at first along the

The Mediterranean is an archaeological treasure store. Here, divers work on the site of a sunken, possibly forgotten, city, using compressed air to jet away sand, while air-filled bags gently lift a new discovery to the surface.

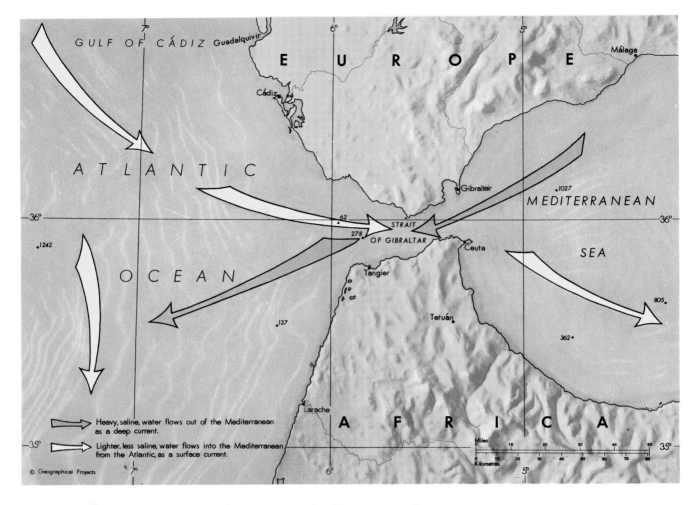

north coast of Africa and then fans out, gradually evaporating so that it becomes heavy with salt and sinks. "Heavy" water at the western end of the Mediterranean piles up against the Strait of Gibraltar lip until it overflows into the Atlantic. Without this flow and counterflow the Mediterranean would become a stagnant, rapidly diminishing pond.

The inward-flowing Atlantic water and outward-flowing Mediterranean water form distinct layers. The outward-flowing water is heavy with dissolved salts and so creeps over the lip, while the lighter Atlantic water flows in over the top of it. It was this phenomenon that the German and Italian submarines exploited to such advantage. Their commanders knew that they could get in and out without using their engines merely by placing their craft in the relevant inward- or outward-flowing current and drifting silently along. They could even escape fingers of sound with which their hunters probed the waters of the Strait. The tracking ships transmitted these ultrasonic pulses through the water in the hope that they would bounce off the hulls of the submarines to give a revealing echo to the listeners above. But the path of sound waves is peculiarly affected by temperature gradients in water, and the counter-flowing streams at the Strait of Gibraltar create a complex structure of such gradients, known as thermoclines. The sound waves could not pass through the boundaries of these gradients but bounced impotently off, making the chances of discovery very slim.

Ugly playthings on a Mediterranean beach—but just a small example of the way in which the cradle of civilization is slowly being choked to death by the unwanted products of industrialized society.

This map shows the narrow entrance to the Mediterranean Sea—without which the sea would be completely landlocked. "Heavy" salt-laden water piles up against the lip at the Strait of Gibraltar until it overflows into the Atlantic. Inward-flowing water from the Atlantic is lighter, and so the ingoing and outgoing streams form two distinct layers.

Engineers have not been slow to recognize the potential offered by this inability of the Mediterranean to sustain itself without the life-giving inflow of water from the Atlantic. "Build a dam across the Strait of Gibraltar," they say, "and within a few decades the level of the Mediterranean will fall to such an extent there will be a difference between the Atlantic and Mediterranean of perhaps a hundred feet." This difference could be utilized to generate power on a vast scale, and would provide great tracts of new land along the Mediterranean coasts. Within a hundred years the level would drop by perhaps as much as 300 feet and it would also be possible to take advantage of the sill that divides the eastern and western Mediterranean across the Strait of Sicily (between Sicily and North Africa). By damming this strait and the Strait of Messina (between Sicily and Italy) the two halves of the Mediterranean could be kept at different levels and even more electric power generated.

MEDITERRANEAN SEA
General Features/Average Surface Currents

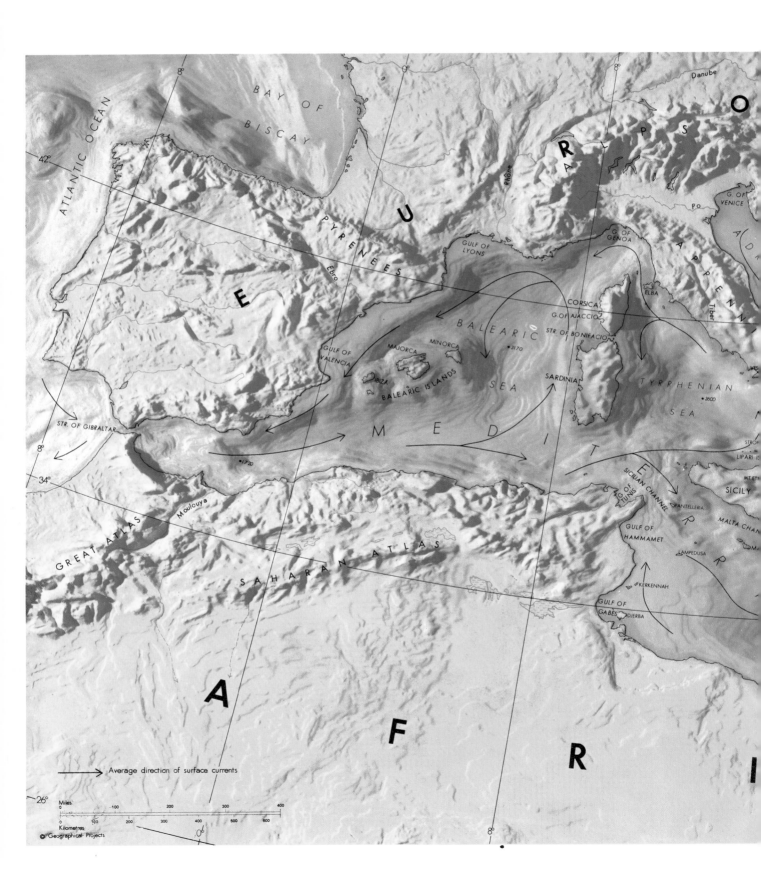

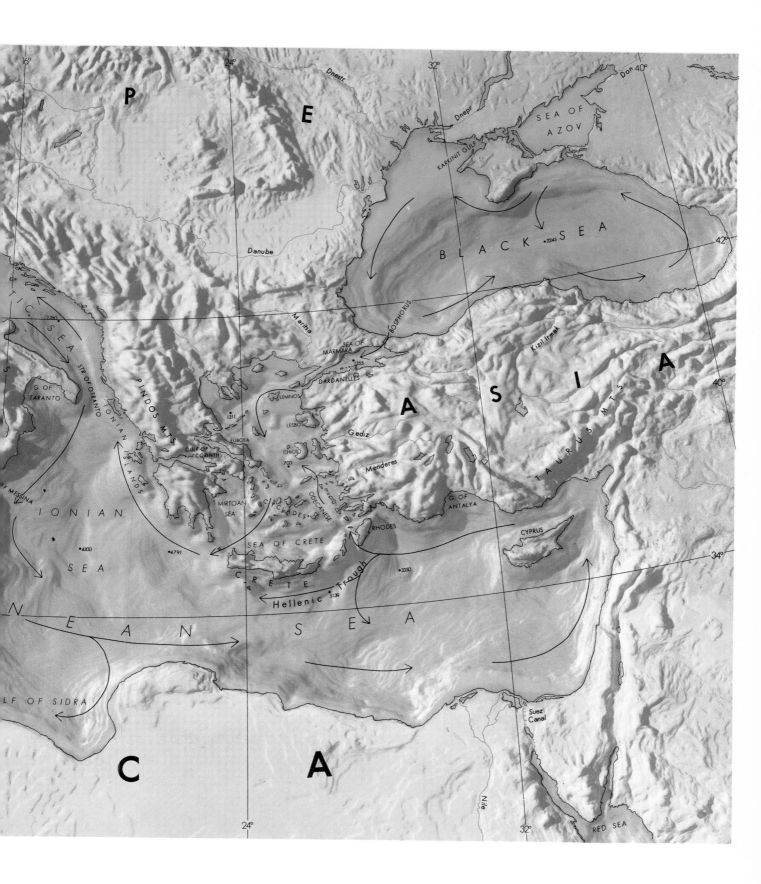

The Mediterranean Sea, showing average surface currents. Because the sea is nearly landlocked, these currents are comparatively sluggish, and the tidal range is only a few inches in most places.

It is unlikely that these ideas will be adopted. As man becomes increasingly aware of the often disproportionate effects his exploitation of the natural environment can have on the overall balance of nature, so he realizes the catastrophes that could be wrought by large-scale interference of the type a Gibraltar dam would represent. Yet if such dams were ever constructed and the level of the Mediterranean could be made to fall by a few hundred feet, what an archaeological treasure trove would be revealed! The entire Mediterranean coastline is littered with the remains of early civilizations. Hundreds of cities and countless wrecks lie beneath the waves, waiting for the combination of diligent research and happy accident to reveal to the archaeologist their wealth of information about previous civilizations.

How did these cities of great seagoing explorers and traders—Phoenicians and Egyptians, Greeks and Romans—come to be consumed in their hundreds by the waves? Even today the answer is not complete, making the Mediterranean a fruitful hunting ground for geologists, oceanographers, and archaeologists alike. Sometimes the interests merge. Nicholas Flemming, a geologist from Britain's Institute of Oceanographic Sciences, took part in an underwater research expedition to survey the city of Apollonia in North Africa in 1958. He became curious about the reasons for the city's submergence and this has led him to conduct a program of research aimed at explaining the changes in the level of the Mediterranean Sea over the centuries. In the course of this he has swum over the sites of more than 100 underwater ruins and a score of wrecks. It would be a dry academician who could investigate such sites for their geological interest alone, and Flemming was soon bitten by the archaeological bug.

Left: A deep-sea diving bell is lowered into the Mediterranean. The divers breathe a helium-oxygen mixture at a pressure equal to the eventual depth the bell will reach. Once at the work site (above), they emerge through a hatch at a depth of 840 feet to carry out a simulated oilfield task. When the work is complete they re-enter the bell and, still under pressure, are winched to the surface and transferred to a decompression chamber.

The sinking of ancient cities is now seen to be one of the consequences of continental drift. The boundary between the African and European plates runs along the Mediterranean, making the entire region unstable, as Mount Etna in Sicily continues to demonstrate. Nearly all the sunken Mediterranean cities were victims of this instability in the Earth's crust: some were consumed quickly by earthquakes or volcanoes—Pompeii is such an example; others were slowly moved over hundreds of years in the great squeeze between the African and European plates. Coastal erosion took care of some of the others; the accumulated weight of sediments from great rivers caused subsidence in still more; rises in sea level itself—the last vestiges of the melting of the Scandinavian and Canadian icecaps—may have contributed. The whole pattern of events is still not fully understood and will provide geologists and archaeologists with many years of rewarding research.

If the Gibraltar dam was ever built, far less pleasant features than underwater cities and ancient wrecks would be revealed. As the Mediterranean began to retreat from the coasts, centuries of human waste would be uncovered. The inhabitants of Cyprus were among the first to use the oceans as a dumping ground for industrial waste: they mined copper to sell to the Phoenicians and tipped the unwanted material into the sea; the Phoenicians did the same when they mined for tin; and ancient Rome's sewage was carried to the sea by the Tiber. Today, some people believe that the Mediterranean, after centuries of such treatment, is dying—choking to death on the by-products of our industrialized world. Many people no longer give the Mediterranean its old Latin tag of "*mare nostrum*" (our sea) but refer to it instead as "*mare monstrum.*" Many beaches that were once the playground of Europe have been closed to the public because of the accumulation of pollution, waste that the small tidal range of the Mediterranean, only a few inches in most places, cannot scour away.

Generally, the seas and oceans of the world are good places in which to dispose of waste. The global circulation system is an efficient dispersant, and seawater itself is a good diluent. The dumping must be planned to take full advantage of this efficiency and the Mediterranean is just not efficient enough to cope with all that is poured into it. For a start, its only link with the global circulatory system is that narrow outlet at the Strait of Gibraltar. In addition, it depends for life-giving oxygen on the cool, continental air masses that flow across three main areas: the Provençal basin, the upper Adriatic Sea, and the Aegean Sea. Oxygen from these air-flows dissolves in the water to maintain life. But these areas are themselves surrounded by heavily populated, industrialized societies so that pollution is at its worst in these crucial waters, with oxygenation and photosynthesis being inhibited by films of oil and other oxygen-consuming waste. As a result, some say that life in the Mediterranean is fading fast. Swiss marine scientist Jacques Piccard predicts that life in the sea will be dead in 25 years unless fast remedial action is taken; French underwater explorer Jacques

Cousteau says the vitality of the sea has declined by 30 to 50 per cent in the past 20 years. Even when one allows that these two underwater pioneers have sometimes overstated their case to emphasize the urgency of anti-pollution measures, there would still seem to be cause for concern.

A massive program of international control is needed to keep the Mediterranean alive. But at the moment, ironically, it seems that the biggest efforts being made are those that seek to extend even further the exploitation of the area. For despite the pollution there are still many parts of the Mediterranean that have the clear, warm water that makes it an ideal area for underwater research. Diving for profit first began in the Mediterranean when ancient Greeks dived to depths of 100 feet to recover sponges. In this century, Jacques Cousteau first experimented with the aqualung off the French Riviera; later he and American pioneers such as Edwin Link carried out their separate experiments in underwater living off the warm Mediterranean shores.

These experiments formed the basis of later experiments that are now leading to man's being able to work, unprotected, at depths of over 1,000 feet in the oceans. Today the Mediterranean plays just as large a part in the development of these underwater work techniques. At Marseilles, for example, is the headquarters of one of the world's largest commercial diving organizations, Comex. In September 1970, in the Bay of Ajaccio off Corsica, three Comex divers carried out some quite complex engineering tasks at a depth of 840 feet, the greatest depth at which a free-swimming diver has ever worked, although others have made quick "bounce" dives to depths greater than 1,000 feet. The experience gained in this experiment, in which the three divers spent eight working days underwater, of which 20 hours was at 840 feet, will quickly be put into commercial practice. Nearly all commercial diving research is today undertaken to satisfy the ever more demanding needs of the offshore oil industry as it moves into deeper and deeper water, and the Comex divers were in fact simulating typical underwater oil exploration work.

One of the areas in which this deep work will be required is the Mediterranean itself. For despite the present pollution crisis, such is the demand for petroleum and natural gas that the Mediterranean looks to be a prime target for the offshore drillers over the next few years. Already there have been extensive gas discoveries in the Adriatic Sea, while oil has been found off Malta, in the Gulf of Lyons, off Amposta and Castellon, Spain, and in the Gulf of Gabès off Tunisia. The Tyrrhenian Sea is also attracting prospectors.

Much of the oil thought to underlie the Mediterranean is in deep water off the continental shelves, which throughout the whole sea have an average width of less than 25 miles. From the shelves, steep continental slopes fall away to abyssal plains, of which the biggest is the 30,000-square-mile Balearic Plain, 9,000 feet below the surface of the Balearic Sea at its deepest point. The Adriatic, Ionian, and Aegean seas all have abyssal plains, while in the Hellenic Trough, which runs south of Crete,

A massive derrick barge in the Adriatic. It is used for lifting sections of oil and gas production platforms into place. Engineering "giants" such as these are being used more and more as the search for offshore oil and gas intensifies in the Mediterranean.

the Mediterranean reaches its greatest depth in a small depression nearly 17,000 feet deep.

The research ship *Glomar Challenger* has already shown that it is possible to drill in such great depths as these, and has in fact carried out a drilling program in some of the deepest parts of the Mediterranean. Now the oil explorers seem set to follow her.

But can the Mediterranean withstand the full impact of yet another major industry? For while properly conducted oil exploration and production itself contributes hardly at all to marine pollution, the prospects for the Mediterranean, should something go wrong, could be bleak indeed. For example, it is estimated that in 1971 as much as 300,000 tons of oil may have been dumped into the Mediterranean by ships cleaning their tanks at sea or at oil terminals. Each gallon of this oil that decomposed in seawater would have depleted the oxygen from 400,000 gallons of water. If something went seriously wrong with an oil production system in the Mediterranean, and thousands of tons of crude oil were added to this already formidable burden on the sea's meagre oxygen resources, could the Mediterranean cope with the strain? A series of major accidents, combined with increased industrial pollution, could prove the final straw. In which case the engineers might just as well build their dams across the straits. The cradle of civilization will have become a cesspool.

10 The North Sea

It is ironical that one of the most vicious areas of the world ocean should contain so many riches—from herring to sand, gravel, oil, and gas.

If one were allocating the number of words in each section of this book on the basis of the extent of the area of world ocean being described, then the North Sea would merit a short sentence. Compared to other parts of the world ocean, the Atlantic or Pacific, for example, it is a puddle. With an area of 220,000 square miles and an average depth of just over 300 feet it is a smear on the earth's crust. Yet the part it plays in the prosperity of the nations that border it—the United Kingdom, Norway, Denmark, Germany, Holland, and Belgium—is just as important today as at any time.

The North Sea is a particularly nasty little puddle. Just 650 miles long and 400 miles at its widest point (between Denmark and the United Kingdom), its frequently mountainous seas and vicious gales seem out of all proportion to its size. And it is only over the past ten years that the full extent of its viciousness has been fully realized. For although European marine scientists have been measuring and recording in the North Sea for decades, it was not until the offshore oil and gas industry came on the scene in the early 1960's that the extent of the vagaries of North Sea conditions were fully appreciated. Even in the southern section, where scientists have been working for more than a century, it was not until the oil industry moved in that the maximum wave height that could be expected was fixed at an incredible 50 feet. This in an area where the total water depth seldom exceeds 150 feet!

Go north and it gets even worse. The North Sea slopes downwards from south to north, so that while the first drilling rigs operated off the Norfolk coast in 88 feet of water, in the northern North Sea wells are being drilled in over 600 feet of water. Here, marine equipment has to be designed to withstand a

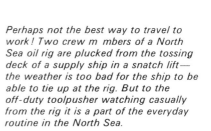

Perhaps not the best way to travel to work! Two crew m mbers of a North Sea oil rig are plucked from the tossing deck of a supply ship in a snatch lift— the weather is too bad for the ship to be able to tie up at the rig. But to the off-duty toolpusher watching casually from the rig it is a part of the everyday routine in the North Sea.

NORTH SEA/BALTIC SEA
General Features

storm that could produce waves up to 100 feet high and winds of 150 miles per hour. Even in a mild winter the waves are over 15 feet high for 20 per cent of the time, water temperature drops to 40°F, and air temperature is frequently below freezing. And unlike areas such as the Gulf of Mexico, where storms can be accurately predicted, the North Sea produces violent weather with only a little warning. A frightening place in which to work.

Sometimes a storm coincides with a high tide to cause devastation in the countries bordering the North Sea. Tides are the result of the sun and moon attracting the oceans towards them. The moon, being closer, plays the greater part in this attraction and pulls towards it the water on the side of the earth facing it. A second bulge of water occurs on the side of the earth facing away from the moon. Here it is the land part of the earth that is closer to the moon, so the land is pulled closer to the moon than the more distant water at the surface. The sea is "left behind," so to speak, and therefore bulges outward slightly from the earth. These two bulges of water move around the earth with the moon, so that two tides occur at any point on the earth's surface every lunar day of 24 hours 50 minutes. The sun also exerts a pull, and when the three bodies are in line—earth, moon, sun—the greatest pull is exerted to form the highest, or spring, tides. When earth, moon, and sun are at right angles a cancelling-out effect gives the smallest tidal range, the neap tides.

Tides are not uniform around the earth because of the complex shape of the ocean basins. The water is pulled first one way and then the other by the sun and moon, and tends to oscillate so that the movement builds up. Tidal range in the open ocean is only a few feet; but as the water in the ocean is directed by tidal forces towards a coastal sea it swills over into the shallower area, where the tidal range can be funnelled or magnified by the land.

This amplification of tidal range occurs in the North Sea, which is funnel-shaped, narrowing to a southern outlet of only 22 miles at the Straits of Dover. So as the Atlantic tide overflows into the northern end of the North Sea the tidal wave moves southwards and is amplified by the funnelling effect. A north wind blowing down the sea also increases the level of water towards the south.

In January 1953, strong winds and high tides combined. A storm swept in from the Atlantic around the Orkney and Shetland islands. It ripped down the east coast of England, combining with a spring tide to raise the water level of the southern North Sea by as much as 10 feet above the predicted tidal level. The low coastal areas of East Anglia and the Thames Estuary were flooded, and more than 300 people were drowned. The water moved in a counter-clockwise direction and hours later smashed into the fragile coast of Holland, breaking more than 400 dikes and causing tremendous loss of life.

Freezing gusts of up to 150 miles per hour; tidal currents of up to 5 miles per hour; waves up to 100 feet high—on this basis the North Sea would seem to be a good place to stay well away from! Yet for its size it is one of the most intensively exploited areas of the world ocean. Thousands of people spend their lives on it; many lose them. The North Sea parts with its riches reluctantly: fishing boats have been known to vanish without trace, or are found drifting, battered, and crewless; oil rigs are ripped from their moorings; some are lost altogether; walls and dikes, built to protect coastal communities from the sea's fury, are ripped apart. But such are the rewards that exploitation of the North Sea is increasing daily.

Fishing, as in so many other areas of the world ocean, is the North Sea's oldest resource industry, and fishermen over the centuries have given the strangely colourful names that so many areas of the sea bear. Some, like Great Silver Pit, Smith's Knoll, and Markham's Hole, can be relatively easily explained as areas where herring were found in great quantities, or where individual skippers had particular success. Others, like Ribs and Trucks Pits, or Tea Kettle Hole, take a little more explaining.

The North Sea teems with fish. Over 55 species of fish and shellfish are caught, the fish ranging in size from tiny sprats to large porbeagles, or mackerel-sharks, the types of shellfish as varied as squids and periwinkles. The whole of the North Sea is utilized for this fishing effort, which produces over three million tons of fish a year valued at well over £100 million (over $250 million). The catching effort is diverse: from remote Scottish ports small lobster boats, often operated by one man, make a good living from a few lobster pots set close to the shore; farther out to sea, carefully organized fleets of Polish and Russian trawlers systematically scour the waters for herring. Fishing vessels of all types and sizes from at least 13 nations fish the North Sea regularly and although the fortunes of individual species may vary, such is the diversity both of species and of catching methods that the overall yield of fish seems unlikely to fall in the foreseeable future, providing pollution is kept within strict limits.

For centuries fishing has been the most important North Sea

The North Sea teems with fish and supports a wide variety of types of fishing boat. These are just three: a Dutch beam trawler (left), which tows a shrimp trawl on each of the side booms; a modern Lowestoft side trawler (centre); and a one-man-operated Scottish lobster boat (right).

In 1953 the North Sea went wild. This is just one of 400 dikes that were breached along the fragile coast of Holland. But the Dutch fought back and are today still reclaiming more and more land from the clutches of the North Sea.

NORTH SEA
Oil and Gas

Part of the deck of a pipe-laying barge. It shows sections of pipe stacked ready to be welded into a string that will then be lowered to the seabed to bring North Sea oil to shore.

This map went out of date as it was being made! Such is the intensity—and the success—of oil exploration in the North Sea that fresh discoveries are being made nearly every month. Just the oilfields shown here are sufficient to provide the United Kingdom with about three quarters of her oil needs by 1980, and to make Norway self-sufficient.

resource industry. But it is fast losing its position as oil begins to flow from the numerous discoveries made in the northern section.

The full effect on Europe of the discovery of oil and gas in the North Sea has yet to be felt. But already it is believed to be so significant that in some quarters it is being called a second Industrial Revolution. Until just a few years ago, north-west Europe seemed to be heading steadily towards an energy crisis. Indigenous natural resources were being depleted at an alarming rate, nuclear power seemed not to offer the promise it had once held, and there was increasing dependence on imported oil supplies from the politically unstable Middle East, whose oil-producing nations, becoming more aware of this dependence, were driving tougher and tougher bargains with Europe.

Now much has changed, and it is amazing how quickly it has all happened. In 1959 a huge land reservoir of natural gas was discovered near the coast of Groningen Province in Holland. This kindled the interest of the international oil industry in what the North Sea might have to offer, and preparations were made to start exploratory drilling. By a combination of clever geological detective work and a share of good luck, the fourth North Sea well to be drilled (by British Petroleum's jack-up rig *Sea Gem*) struck gas. That was at the end of 1965. Further strikes followed, with the result that today over 90 per cent of Britain's gas comes from the southern North Sea and millions of British homes have been converted to accept the new fuel.

After the first wave of successful gas strikes, exploratory activity slackened off, and during 1970 some of the rigs began to move away to areas like the Far East and West Africa, where the prospects for discovering oil were thought to be brighter. Then, in July 1970, the Phillips Petroleum Company, working in the south-west corner of the Norwegian sector of the North Sea, close to the boundary lines of the Danish and British sectors, struck oil. This was the giant Ekofisk field, which is estimated to contain enough oil to make Norway entirely self-sufficient in petroleum products until the end of the century.

The Ekofisk discovery sparked off a flurry of exploratory drilling, and other discoveries followed in quick succession: BP's massive Forties Field, 110 miles north-east of Aberdeen; Shell's Auk field, in the British sector just across from Ekofisk; and the Shell Brent field, 100 miles off the Shetlands. These and other discoveries will have a marked effect on the economies of north-west Europe: the UK, for instance, will have over half her oil requirements met by the North Sea by 1980; Norway will be exporting surplus oil. A further benefit is the massive shore-based activity required to back up marine operations. In Scotland, for example, where until 1971 there were serious unemployment problems, thousands of new jobs are being created over an area stretching from Edinburgh to the Shetland Islands as sites are built to construct production platforms, to prepare underwater pipelines, to provide bases for ships and supplies, and to meet the vast demand for equipment and services.

The benefits are great. So are the costs of bringing oil from the deep waters of the northern North Sea: the development of a major field costs as much as a space shot to the moon. And the technological problems are immense, too. To produce the oil from the Forties Field, the biggest-ever fixed platform is being constructed in Scotland. The main section of the structure (the jacket) will be 475 feet high; with the decks, living quarters, derricks, and other equipment built onto this, the top of the structure will be 800 feet above the seabed! The installation of this structure in the wild waters of the North Sea, scheduled for late 1974, will be one of the greatest offshore engineering feats ever attempted.

Each North Sea oilfield poses a fresh set of problems, and every aspect of marine work—from offshore construction to diving, from ship operation to environmental monitoring—is at present being strained to its technological limits. For if men and equipment can cope with the North Sea, then they can cope with any other offshore oil area in the world. The North Sea will continue to be the proving ground for men, materials, and equipment for many years to come.

Fisheries, oil and gas, and minerals—all three of these great ocean resource industries are represented in the North Sea. But in the case of minerals there is little glamour attached to the industry. For in the North Sea men go hunting not for manganese nodules or seabed diamond deposits, but for lowly sand and gravel—marine aggregates.

As the pace of building quickens in industrialized nations like the UK, the demand for building materials, especially cement, shows a corresponding growth. Transport is the biggest

A group of Dutch companies, led by the Bos Kalis Westminster Group, have proposed building artificial islands in the North Sea to act as oil tanker terminals, as sites for power stations or airports, or as waste-processing stations. They would be built with the aid of the type of dredger shown in the foreground. This trails a pipe that sucks up sand and gravel and pumps it into the vessel's hoppers.

single cost in aggregate production, and so on land the source of supply needs to be as close as possible to the building site, whether this be an office block or a new road. But in many areas, such as London, either there are insufficient stocks of aggregate, or the very land from which it could be produced is required for building purposes. So it becomes economical to exploit marine aggregates close to the centre of demand.

The United Kingdom leads the world in marine aggregate production with over 11 million tons landed in 1971. The main tool of the producers is the trailing suction hopper dredger. These craft, the largest of which are over 300 feet long and can carry up to 8,000 tons of aggregate, have a pipe hinged at the front of the vessel, the free end of which is lowered perhaps over 100 feet to the seabed. Powerful pumps on board the ship suck up the sand and gravel. The latest dredgers have grading devices to separate the sand from the gravel on board the ship.

Dredgers also play a large part in the efforts of countries like Holland to reduce the 220,000-square-mile area of the North Sea. For centuries the Dutch have patiently driven back the sea from their shores, reclaiming vast areas of land for agriculture and building. These projects require dredging on a massive scale: for example, when the new Europoort complex was created at Rotterdam some 400 million cubic yards of material was moved in, to provide 20,000 acres of reclaimed land. As other nations became more and more crowded, Dutch methods of creating new land from the sea will be increasingly adopted. In England, for example, London's third airport may be built on reclaimed land in the Thames Estuary.

Other proposals go still further. Holland's Bos Kalis Westminster Dredging Group plans to build artificial islands in the middle of the North Sea. These could be used as oil terminals, for power stations or airports, as deepwater harbours, or as centres for waste processing and controlled disposal. In its latter role the island would process domestic waste, chemical

NORTH SEA/BALTIC SEA
Tides/Pollution

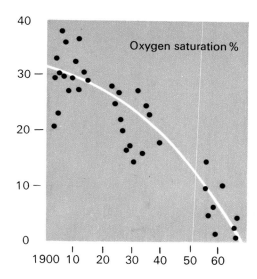

North Sea and Baltic tides and the main sources of Baltic pollution are shown on this map. The red lines represent tidal range at intervals of two feet. The black lines join points that have high water simultaneously at successive lunar hours.

The slow but devastating effect of pollution. Dissolved oxygen content at a position in the northern Baltic at depths of 450–500 feet has been measured since 1900. This graph shows only too dramatically how the amount of oxygen in the water has decreased. If this trend continues, the Baltic Sea will soon become a marine desert.

waste materials, scrap metal and so on, which would be transported from surrounding nations by ship. The energy obtained from incineration would power an air separation plant for the production of nitrogen, oxygen, and hydrogen, and for the liquefaction of natural gas. It could also be used to power a distillation plant producing fresh water.

Projects such as this are going to become increasingly necessary for heavily industrialized nations like those bordering the North Sea. For although the resource industries of the sea themselves add little to its pollution, the communities to whose prosperity they contribute will need to slow down the rate at which they use not only the North Sea but all the world ocean as a great dumping ground. International agreement has now been reached to limit the deliberate dumping of by-products of the chemical and plastics industries—substances like chlorinated hydrocarbons and bromide compounds — in north European waters, but considerably more effort will have to be applied to limit the amounts of domestic and industrial waste that are poured into European rivers and which eventually pollute the seas.

A dramatic example of the toll that pollution is taking in the seas of Europe is supplied by the graph on this page. In 1900 a monitoring device for recording the amount of dissolved oxygen—essential to marine life—was installed in the Baltic Sea at a depth of about 350 feet. The level of dissolved oxygen was recorded at that site over the next 70 years. That slumping line tells its own story: if it continues, the Baltic will become a marine desert, killed by man's waste.

The oceans and seas of the world hold out the greatest single hope for man's survival and well-being on this planet. They also hold the key to his destruction. The choice is ours: exploit the resources of the world ocean rationally and with careful regard to the overall oceanic system; or grab resources indiscriminately while continuing to use the oceans as a trash pit. The story of the whale and that dissolved oxygen graph are pointers to the consequences of following the second course.

Index

References in *italics* are to illustrations or captions to illustrations. References in **bold** are to map keys or captions.

abyssal plains, 8, 14, *14–15*; in Gulf of Mexico, 94; in Mediterranean, 110
Adriatic Sea, oil and gas in, 110
African continental plate, 22, **22**, 63, 109
Agulhas Current, 57
aircraft, location of tuna shoals by, *40–1*, *56–7*, 58
Alaska: king-crab fishery off, 41; transport of oil from North Slope in, 70–1
albacore (tuna), *41*
Aleutian Trench, earthquake in, 34
American continental plate, **20**, 22, **46**
anchovy fisheries: in Caribbean and Gulf of Mexico, **96–7**; off Peru (anchoveta), *36–7*, **37**, 40–1, 57
Andes Mountains, world's tallest range, 37
Antarctic Circumpolar Current, 72, *72–3*
Antarctic Convergence, 74
Antarctica: landscape of, *74*; percentage of world's ice and snow in, *74*
Arabia, upwelling off coast of, 57
archaeology, underwater: in Mediterranean, *102–3*, *108–9*
Arctic Ocean, 64, 66, **67**; conditions in, *64–5*, *68–9*; depth of, 66; oil in, 68, 69–70; possible icebreaker and submarine routes across, 70–1, **70–1**
Atlantic Ocean, *18*; area and depth of, 16; fisheries of, 25, **26–7**, 28, *28–9*, 30–1; Gulf Stream in, 23–4; islands in, 19; Mid-Atlantic Ridge in, 19–23; Sargasso Sea in, 24–5
Atlantis Deep, Red Sea, 63
Azores-Gibraltar Ridge, **22**

Baltic Sea, **114**; pollution of, **122–3**, 123; tides in, **122–3**
barracuda, 87
basins, of Atlantic, 19
beach-seining, *48–9*
Beagle, H.M.S., 36
Ben Franklin submarine, *25*
Bering Sea, 87; fishing in, *86–7*
bigeye tuna, *40*
blue whales, *78*, 80, 82, 83
blue whiting, in Atlantic, **26–7**; not yet exploited, 31

bluefin tuna, *40*
buoys, possibility of automatic oceanographic, *56–7*, 58

California: Antarctic icebergs as possible source of fresh water for, *76–7*, **76–7**; oil leak on coast of, 46, *46–7*
Canada, fishing fleets of, 87
canoes, dugout, fishing from, *48–9*, 58
Cape Horn, 72
Caribbean Sea, 92, 94, **96–7**; depth of, 92; fisheries of, 28, 95, **96–7**; oil and gas in, **96–7**
Cariaco Trench, 94
catch quota system, for some North Atlantic fishing grounds, 31
Cayman Trench, 92
cephalopods, fishing for, 25, 99
Challenger, H.M.S., 48
cobalt, on Pacific floor, 42
cod: in Atlantic, 25, **26–7**, *29*, 30, 31; in Far Eastern Seas, 87, *87*; ling (or King) variety of, *29*, 31
Comex diving organization, Marseilles, 110
computers, to process data from aircraft, satellites, and automatic buoys, *56–7*, 58
continental drift, 22; sinking of ancient cities as consequence of, 109
continental shelves, 14, *14–15*, 25; in Arctic, 68; in Atlantic, 16, 19, 25; in Caribbean and Gulf of Mexico, 94, 98, 100; in Indian Ocean, 48; in Japan Sea, 88, 91; in Mediterranean, 110; as percentage of earth's surface, *14*; in Southern Ocean, 76
continental slope, 14, *14–15*, 94
Convergences, Antarctic and Subtropical, 74
copper: at bottom of Red Sea, **62**, 63; in Cariaco Trench, 94; mining of, in ancient Cyprus, 109; on Pacific floor, 42
coral atolls and reefs, *32–3*, *36–7*, *39*
Coriolis Effect (of earth's rotation), 52, 57
Cousteau, Jacques, underwater explorer, 109–10
Cuba: fisheries export agency of (Caribex), 98–9; fishing fleets of, 95, 98
currents: Antarctic Circumpolar, 72; earth's rotation and, 37, 52; of Indian Ocean, monsoon winds and, **51**, 52, 57; of Mediterranean, **106–7**; meter for recording, *24*; at Strait of Gibraltar, 102, 104, **104–5**; of world oceans, *12–13*; *see also* Gulf Stream

dab, in Atlantic, **26–7**
Darwin, Charles, and coral atolls, 36
Davis Strait, oil prospecting in, 68, 69
Deepsea Miner, 42
derrick barge, *110–11*
diesel engines, in fishing ships, 28
diving: aqualung, 110; archaeological, *102–3*; in capsule at atmospheric pressure, *100–1*; off coral reef, *32–3*; in deep-diving bell under pressure, *108–9*; for servicing underwater oilfields, *60–1*, 63; for sponges, 110
dogfish, in Atlantic, *29*
dolphins, 81; associated with shoals of tuna, *40–1*
Drake Passage, 72, *72–3*
drift-netting of herring, 30–1
drilling for oil and gas, *see* oil drilling
Dubai Sheikdom, *58–9*, 61

earth: rotation of, and ocean currents, 37; word used both for planet, and for soil, 8
earthquakes, 20; in California, 43, **46**; in Mediterranean, 109; undersea, causing tsunamis, 34
East Germany, fishing fleet of, 28
echo-sounder, *see* sonar
eels, life history of, 24–5, *25*, **25**
Ekofisk oilfield, Norwegian North Sea, 120
El Niño current, off Peru, 40–1
elephant seals, *82–3*
energy, prospective shortage of sources of, 46, 119
England, early fisheries of, 28
environment, standards of life and of, *46–7*
Equatorial Counter-Current, 57
Equatorial Currents, North and South, 52, 57
Eurasian continental plate, 22, **22**, 63, 109

Fairtry I, first stern-fishing freezer trawler, 30

125

Falkland Islands, 28
Far Eastern Seas, 84, *84–5*, **86**, 87; fisheries of, 87, *87*; Japan and, 88, 91; oil in, **86**, 88
Fateh oilfield, Persian Gulf, 58–9, 61
fin whales, *78*, 82
fish: of Atlantic, **26–7**, *29*; of Caribbean and Gulf of Mexico, **96–7**; decline of stocks of, 31; demersal and pelagic, 25; in Marianas Trench, 32; species of, not yet exploited, 31
fisheries, *see under individual oceans and seas*
fish farming: in Cuba, 98; in Japan, 91
fishing: on continental shelves, *14–15*; different methods of, *28*, *40*, *48–9*
fish meal, as animal feeding-stuff, *37*, 40, 41
flatfish: in Atlantic, 25, **26–7**; in Far Eastern Seas, 87, *87*
Flemming, Nicholas, underwater geologist and archaeologist, 108
floor of ocean: mineral nodules on, 42; sonar charts of, 23
Food and Agriculture Organization of United Nations, 57–8, 99
food chain, supporting fish, 40
Forties oilfield, North Sea, 120
Fram, *66*; Arctic voyage of, 66
Franklin, Benjamin, has Gulf Stream charted, 16, 23
freezing of fish on board ship, 30, *30–1*
fulmar, *82–3*

General Dynamics Corporation, 69, 70
Germany: East, fishing fleet of, 28; redfish eaten in, 31; West, deep-sea mining group in, 43
Gibraltar, Strait of, 109; inflow and outflow of water at, 102, 104, **104–5**; possibility of dam at, 105
glaciers, icebergs from, 69
Glomar Challenger, deep-water drilling ship, *21*; in Atlantic, 22; in Gulf of Mexico, 94–5, 101; in Mediterranean, 111
Gloria fault, along Azores-Gibraltar ridge, **22**
GLORIA (Geological Long-Range Inclined Asdic) sonar instrument, 22, *22*, 23
gray whales, ban on hunting of, 82
Greenland, glaciers of, 69
grouper: in Caribbean and Gulf of Mexico, **96–7**, 99; in Far Eastern Seas, 87
Gulf Stream, 16, 23–4, *25*, 57
Guyot, Arnold, discoverer of seamounts (guyots), 36
gyres: of Northern Hemisphere, 52, 57; of Southern Hemisphere, 57

haddock, in Atlantic, 25, **26–7**, *29*

hake: in Atlantic, 25, **26–7**, 41; in Caribbean and Gulf of Mexico, **96–7**; on Patagonian shelf, 28, **30**, *31*
halibut: in Atlantic, 25, **26–7**, *29*; in Far Eastern Seas, 87, *87*
harpoon cannon, used in whaling, 81, 83
Hawaii: devastation by tsunami in, 34, 46; volcanic origin of, 37
Hellenic Trough, 110–11
herring: in Atlantic, 25, **26–7**, *29*; in Caribbean, 99; off Falkland Islands, 28; in Far Eastern Seas, 87, *87*; in North Sea, 116
Hokkaido, devastation by tsunami in, 34
Holland: floods of 1953 in, 115, *117*; natural gas in, 119; reclamation of land in, 121
Holt, Sidney, of FAO, 58
Honolulu, central base of Tsunami Warning System in, *35*, 36
hopper dredgers, 121, *121*
"hot spots," in depths of Red Sea, 63
Hughes Tool Company, 43
Hult, John, of Rand Corporation, 77
humpback whales, *78*, 82
hurricanes, *92*, 100

ice: in Antarctica, 74, 76; in Arctic Ocean, 69; in Bering Sea, 87
icebergs: Antarctic, as possible source of fresh water, 76–7, *76–7*; in Arctic Ocean, 64, 69; in North Atlantic, 64, 69; in Southern Ocean, 76
Iceland, 19; fishing grounds of, 28; on north end of Mid-Atlantic Ridge, *16–17*
Indian Ocean, **50–1**; area and depth of, 48; currents of, **51**, 52, 57; fisheries of, *48–9*, *52–3*, 57–8; phytoplankton of, *54–5*; ridges on floor of, 52
Indonesian Archipelago, 84
International Indian Ocean Expedition (1960–5), 48, 52, 57
International Nickel Company, 43
International Observer Scheme, for whaling, 82
International Whaling Commission, 82
International Whaling Convention, 82
iron: at bottom of Red Sea, 63; sands yielding, off Japan, 88; in water of Cariaco Trench, 94
islands: of Atlantic, 19; continental and oceanic, 19; man-made, 61, 88, 91; proposed man-made, for North Sea, 120–1, *120*, 121, 123; volcanic, *16–17*, 19, 36–7

Japan: deep-sea mining planned by, 43; fish in diet of, 41, 88; fishing fleets of, 28, **40**, 41–2, *87*, 88; shipbuilding in, 88, *88–9*, 91; whaling by, 82
Japan Sea, 84, 88
Java Sea, 87
junks, Chinese, *84–5*

Kaufman, Raymond, of Deepsea Ventures, Inc., 42–3
Kharg Island, man-made for loading giant oil tankers, 61
king crabs, 41
Kodiak, Alaska, devastation by tsunami at, *34–5*
krill, food of baleen whales, 80; as possible human food, 83; in Southern Ocean, **79**
Kuroshio (Japan) Current, 57

Las Palmas, Canary Islands, as trans-shipment port for fishing fleets, 28, 98
Ling (variety of cod), 31; in Atlantic, 29
lobster fisheries: off Cuba, 98, 99; off Scotland, 116, *117*
Lockheed Aircraft Corporation, 43
Lockheed Petroleum Services, 101
longlining, **40**, 41–2

mackerel, in Atlantic, 25, **26–7**, *29*
magnetism, dating rocks by, 22
magnetometer, *20*
manganese: at bottom of Red Sea, **62**, 63; in nodules on Pacific floor, 42–3, *42–3*
Manhattan, tanker, journeys through North-West Passage, *68–9*, **68–9**, 70
Maracaibo Lake, Venezuela, drilling for oil in, *98–9*, 99
Marianas Trench, Pacific, 32
marlin, 41
Maury, Lt, M.F., USN, 23
Mediterranean Sea, **106–7**; archaeology of, *102–3*, *108–9*; boundary between continental plates in, 109; depth of, 102; oil and gas in, 110–11; pollution of, *104–5*, 109–10
menhaden, in Caribbean and Gulf of Mexico, **96–7**
metals: at bottom of Red Sea, **62**, 63; on Pacific floor, 42–3
Mexico, Gulf of, 94–5, **96–7**; fisheries of, *94–5*, **96–7**; oil and gas in, **96–7**, 99, 100–1, *100–1*
Mid-Oceanic Ridge, 19; in Atlantic (Mid-Atlantic Ridge), 19, 22, 66; in Indian Ocean, 48, 52; in Pacific, 43
migration: of eels, 24–5, *25*; of whales, 80–1
Monsoon Drifts, North-east and South-west, 57

monsoon winds: in Far Eastern Seas, 87; in Indian Ocean, 51, 57
moon, and tides, 115
mountain ranges, at junctions of continental plates, 20; world's tallest (Andes), 37
mountains, from seabed, 8, *14–15*
mullet, in Caribbean and Gulf of Mexico, **96–7**

n

Nansen, Fridtjof, 66
Nautilus, nuclear submarine, under Arctic ice, 70, *71*
Newfoundland, Grand Banks of, 28
nickel, on Pacific floor, 42
Ninetyeast (Carpenter) Ridge, in Indian Ocean, 52
North Sea, **114**; area and depth of, 112; conditions in, 112, 115; fisheries of, 28, 116; oil and gas in, 112, *112–13*, **118**, 119–21; pollution of, **122–3**, 123; tides in, **122–3**
Norway: ice-free ports of, 23; and North Sea oil, 120; whaling by, 82

o

observation tower, underwater (Japan), *90–1*
oceanographic expeditions, 19; Indian, 48, 52, 57
Oceanographic Sciences, UK Institute of, 23, 108
Oceanology International Conference (1972), 42
oil (petroleum): pollution of water by, 46, *46–7*, 109, 111; for sources of, see under individual oceans and seas
oil (from whales), 83
oil drilling, underwater: derrick barge for, *110–11*; different types of equipment for, 99–101; fire on platform for, *92–3*; ice-resistant rig for, *66*; ships for, 23; transport of crew from supply ship to rig for, 112–13
oozes of ocean floor: calcareous and diatomaceous, 43, **44–5**; siliceous, **44–5**
Ostrander, Neill, of Rand Corporation, 77
outboard motors, supplied by UN agency, *48–9*, 58
oxygen, dissolved in water of Baltic (1900–70), 123, *123*

p

Pacific Ocean, **38–9**; area and depth of, 32, 34; coral atolls and reefs in, *32–3*, 33, 36–7, 39; fisheries of, 37, 40–2; oil leak into, *46–7*; sediments in, 42, **44–5**; tsunamis in, 34–6
Pacific plate, 43, **46**
Patagonian Shelf, 28, **30**
penguins, Adélie, *82–3*
Persian Gulf: area and depth of, 61; oil wells and pipelines in, *56–7*, 58; underwater oilfields in, *60–1*, 61, 63; underwater oil-storage tanks in, *58–9*, 61
Peru, fisheries of, *36–7*, 37, **37**, 40–1, 57
Peru-Chile Trench, 37
Peru (or Humboldt) Current, 37, 40, 74
Philippine Sea, 84
phytoplankton: of Indian Ocean, *54–5*; upwelling water and, 37, 40; of world oceans, **55**
Piccard, Auguste, bathyscaphe builder, 32
Piccard, Jacques, marine scientist, 32, 109
pipe-laying barge, *119*
plaice, in Atlantic, 25, **26–7**, *29*
plankton, see phytoplankton
plates of earth's crust, **20–1**, 21–2, **22**, 43, **46**, 63, 109; rate of movement of, 22
Poland, fishing fleets of, 28, 116
pollution: of Mediterranean, 109–10; of North Sea and Baltic, **122–3**, 133; by oil, 46, *46–7*, 109, 111
Pompeii, 109
Portugal, early fisheries of, 28
poultry, Peru anchovy catch affects UK price for, 36–7
power blocks, for purse seining, 31
protein: countries short of, *48–9*, **55**, 57, 58; fish in Japan's supply of, 41, 88
purse seining, *28*, 31, *35–6*, *40–1*

r

rays (skate, etc.), in Atlantic, **26–7**
Rastrelliger kanagusta (of mackerel group), *53*
red clay, on Pacific floor, 43, **44–5**
Red Sea, **62**, 63
redfish (ocean perch, sea bream): in Atlantic, 25, **26–7**, *29*, 31; in Bering Sea, 87
ridges of ocean floor: aseismic, 52; Mid-Oceanic, 19, 22, 43, 48, 52, 66
rift valley, along Mid-Oceanic Ridge, 19, 21
right whales, ban on hunting of, 82
Rotterdam, Europoort at, 121

s

saithe (coley), in Atlantic, **26–7**, *29*
salmon: off Canada, 41; in Far Eastern Seas, 87, *87*
San Andreas fault, 43, **46**

sand: for building materials, from North Sea, 120–1; iron-containing, off Japan, 88
sardines, 25; in Atlantic, 41; in Caribbean, 98; in Far Eastern Seas, 87; in Indian Ocean, *53*; Venezuelan, **96–7**
Sargasso Sea, 24, **24**
Sargassum, tropical seaweed, 24
satellites, observations of oceans by, *56–7*, 58
Schmidt, Johannes, 24
Scotland: effects of North Sea oil in, 120; lobster fishing from, 116, *117*
Sea Gem, jack-up oil-drilling rig, 119
seamounts (guyots), in Pacific, 36
sea trout, in Caribbean, 99
sea water, chemical composition of, 14; at bottom of Red Sea, 63; in Cariaco Trench, 94; in upwelling areas, 37
seas, adjacent (or marginal) and mediterranean, 84
sediments: overlying mid-oceanic rift valley, 22; in Pacific, **44–5**; projects for using, 43
sei whales, 82
seismic sea waves, *see* tsunamis
seismic stations, around Pacific, **35**, 36
shrimps, fisheries for: in Caribbean and Gulf of Mexico, 28, 41, *94–5*, 95, **96–7**, 99; in Far Eastern Seas, 87
silver, at bottom of Red Sea, 63
skipjack tuna, *41*
sleds, underwater, used by divers servicing oil installations, *60–1*, 63
smelt, great silver (argentine): in Atlantic, **26–7**; not yet exploited, 31
snapper: in Caribbean and Gulf of Mexico, **96–7**, 99; in Far Eastern Seas, 87
sole, in Atlantic, 25, **26–7**
Somaliland, upwelling off coast of, 57
sonar, 16; for fishing, 31, *52–3*, 58; for recording ocean floor, 22, **22**, 23
South America, upwelling on west coast of, 37, 40
South Georgia, whaling station on, 81
Southern Ocean, 72, *72–3*, 74, **75**; animals of, *82–3*; ice in, 74, 76–7; whales in, *78–9*, 79, **79**, 80, *80*, 81–3
Spain, fishing fleet of, 28
sperm whales, 81, 83
sponges, 98, 110
steam power in ships, extends fishing grounds, 28
storage, underwater, *58–9*, 61, 91
straw, used in effort to absorb oil leak, *46–7*
submarines: entry of, into Mediterranean during World War II, 102, 104; under Arctic ice, 70–1, *70–1*
Subtropical Convergence, 74
Sunda Shelf, 87, 88
Surtsey, new volcanic island, *16–17*, 19
swordfish: in Caribbean, 99; in Pacific, 41

t

tarpon, *94–5*
telegraph cable, Atlantic, 19
Telegraph Plateau, mid-Atlantic, 19

127

television, underwater, *42*
Tenneco Oil Company, 43
terrigenous deposits, on Pacific floor, **44-5**
thermoclines, in water at Strait of Gibraltar, 104
tidal range, 115; in Mediterranean, **106-7**, 109; in North Sea, 115
tide stations, around Pacific, **35**, 36
tides, 115; coincidence of storms with high (North Sea), 115; in North Sea and Baltic, **112-3**
trade winds, 37, 52, 72
trawler, in Arctic night, *64-5*
trawling: aimed, *52-3*, 58; beam-, *116*; midwater and bottom, *28*, 31; side-, 30, *117*; stern-, 30
trenches in seabed, 8, *14-15*, 20, 21; in Caribbean, 92; in Japan Sea, 84; in Pacific, 32, 34
Trieste bathyscaphe, 32
Tsunami Warning System, **35**, 36
tsunamis, 34-6, *35*
tugs: icebreaker, *66*; to tow icebergs from Antarctic, *76-7*; to keep icebergs off drilling ships, 69
tuna: in Atlantic, 28; in Caribbean and Gulf of Mexico, **96-7**, 99; in Far Eastern Seas, 87; in Pacific, *40-1*, 41
turbot, in Atlantic, 25, **26-7**
Typhoon, drilling ship, in Arctic, 69

United Nations: aid to fishermen from, *48-9*, 58; and Caribbean fisheries, 99; Human Environment Conference called by, 82

United States of America: Caribbean bases of seafood companies of, 99; Coast and Geodetic Survey of, 36; fishing fleets of, *87*, 99; import of whale products banned by, 80, 83
upwelling, of water full of nutrients from ocean depths, 37, 40, 57
USSR: and Cuban fisheries, 95; fishing fleets of, 28, *87*, 116; plans oil production off Siberia, 69; plans use of krill for human food, 83; whaling by, 82

volcanic activity: in depths of Red Sea, 63; in Mediterranean, 109; in rift valley between crests of Mid-Oceanic Ridge, *16-17*, 19, 21; undersea, causing tsunamis, 34
volcanoes: coral atolls on, 36, *39*; on Pacific floor, 36-7

Walsh, Don, USN, 32
waste products: dumped into oceans, 14, 123; in Mediterranean, *104-5*, 109, 111; scheme for processing, 121, 123
water, percentage of world's fresh, in Antarctic ice, 76; *see also* sea water

waves: in hurricanes, 100; in North Sea, 112, 115; in open ocean, 34; on shorelines, 35; of tsunamis, 34, 35
Wegener, Alfred, originator of theory of continental drift, 20
whalebone, 83
whales, *78*; baleen and toothed, 80, 81; in Bering Sea, 87
whaling, 41, 80, 81-3; areas of, **79**
whiting, in Atlantic, **26-7**
winds: and currents, 52, 72; and waves, 34
world oceans, **10-11**; area of, compared with that of land, 8, 14, *14-15*; man's divisions of, *8-9*; phytoplankton of, **55**; resources of, 15; surface currents of, **12-13**
World War II: in North Atlantic, 16; entry of German submarines into Mediterranean during, 102, 104

Yellowfin tuna, *41*
yucatan Strait, 92

Zakum underwater oilfield, Persian Gulf, *60-1*, 61, 63
zinc, at bottom of Red Sea, **62**, 63

Acknowledgments:

We are gratefully indebted to the following who kindly supplied original pictures from which we based our own illustrations:

Aeromarine Photographic Ltd. 121(B); B.P./Total 60; Purnell's *Encyclopaedia of Animal Life* (artist Malcolm McGregor), BPC (Phoebus) Publishing Ltd. 3, 40-41(T); ed. P. J. Herring and M. R. Clarke, *Deep Oceans*, Arthur Barker Ltd. 40(B); Barringer Research, Harpenden 20(B); after L. Bertin, *Eels* 24(T); The British Gas Corporation 113; Camera Press 85, (Hitachi Zosen) 90, (B. Logan) 73; Chicago Bridge & Iron Company 59; Photo A. Tocco, Comex 108-109; Photo B. J. Nixon, Deepsea Ventures, Inc. 42; ELAC Electroacustic GmbH, Kiel 53(T); Food and Agriculture Organization of the United Nations 49, 54-55; General Dynamics 66, 71(L); Global Marine Inc. 21; Photo Ernest Graystone, Lowestoft 116-117(B); Photo R. J. Griffith 80; Grumman 25(B); Humble Oil Company 68-69(B); Institute of Oceanographic Sciences, Wormley, Godalming, Surrey 23(R); International Council for the Exploration of the Sea, Denmark 123(R); Ishikawajima-Harima Heavy Industries Co., Ltd. 89; Photo Ford Jenkins, A.I.I.P., Lowestoft 119; Lockheed 100, 101; Micoperi s.p.a., Milan 111; Photo Josef Muench 47; Photo Don Mackay/*Newsweek*, August 21, 1972 76-77; Novosti 87; The Plessey Company Limited 24(B); after *The Russian Atlas of Antarctica* 79(R); *Science Journal*, December, 1967, Vol. 3, No. 12 35(B); Shell International Petroleum Co. Ltd., 98; United Press International (U.K.) Ltd. 93; U.S. Bureau of Commercial Fisheries 56(T); Photo George Plafker/U.S. Geological Survey 35(T); Photo Peter Waugh 30-31(B); Bos Kalis Westminster Dredging Group 120(T); Whale Research Unit, Institute of Oceanographic Sciences 79(BL); White Fish Authority, Edinburgh 26-27.